Powering the World's Airliners

Powering the World's Airliners

Engine Developments from the Propeller to the Jet Age

Reiner Decher

AIR WORLD

First published in Great Britain in 2020 by
Pen & Sword Air World
An imprint of
Pen & Sword Books Ltd
Yorkshire – Philadelphia

Copyright © Reiner Decher 2020

ISBN 978 1 52675 914 6

The right of Reiner Decher to be identified as Author of this work has been asserted by him in accordance with the Copyright, Designs and Patents Act 1988.

A CIP catalogue record for this book is
available from the British Library.

All rights reserved. No part of this book may be reproduced or transmitted in any form or by any means, electronic or mechanical including photocopying, recording or by any information storage and retrieval system, without permission from the Publisher in writing.

Printed and bound by Replika
Press Pvt Ltd, India.

Pen & Sword Books Limited incorporates the imprints of Atlas, Archaeology, Aviation, Discovery, Family History, Fiction, History, Maritime, Military, Military Classics, Politics, Select, Transport, True Crime, Air World, Frontline Publishing, Leo Cooper, Remember When, Seaforth Publishing, The Praetorian Press, Wharncliffe Local History, Wharncliffe Transport, Wharncliffe True Crime and White Owl.

For a complete list of Pen & Sword titles please contact

PEN & SWORD BOOKS LIMITED
47 Church Street, Barnsley, South Yorkshire, S70 2AS, England
E-mail: enquiries@pen-and-sword.co.uk
Website: www.pen-and-sword.co.uk

Or

PEN AND SWORD BOOKS
1950 Lawrence Rd, Havertown, PA 19083, USA
E-mail: Uspen-and-sword@casematepublishers.com
Website: www.penandswordbooks.com

Contents

Dedication — vi
Glossary of Abbreviations — vii
Dramatis Personae: Pictorial People Index — ix
Acknowledgments — xiv
Objectives — xv

Chapter 1	We are Flying!	1
Chapter 2	Aviation is All the Rage	10
Chapter 3	The 1930s: Real Transport Aeroplanes	42
Chapter 4	Engines for the War Effort	72
Chapter 5	Cold War Engines and Post (Hot) War Airliners	99
Chapter 6	Jet Airliners	125
Chapter 7	The Turbofan Engine	147
Chapter 8	How to Buy an Airliner	187

Epilogue — 204
Appendix: How Does the Jet Engine Work? — 206
Bibliography — 211
Index — 213

Dedication

As a young man, I steered into aeronautical engineering as a field of study and later spent a lifetime teaching other young men and women what I knew and understood. My father, Siegfried Decher, was active in the field and it seems as though he had an almost invisible hand in guiding what I was to do in life. He approved of my choices and held back from an active role in them. In my studies I learned from Jack L. Kerrebrock at the Massachusetts Institute of Technology who taught me to love the magic of fluid flow and its challenges. Later, as I started my career as a teacher, I was privileged to have as a mentor my colleague Gordon C. Oates at the University of Washington, from whom I learned to love thermodynamics. To them I owe much. Among the tasks I undertook in my professional life, I applied my passion, albeit unsuccessfully, to the challenge of finding an engine configuration that would make a supersonic transport plane practical. For that opportunity, I thank Gary C. Paynter of The Boeing Company's Propulsion Research Group for much stimulating discussion and collaborative labour.

I am now sufficiently aged to have enjoyed the company of a number of the engine pioneers and aeroplane-design leaders mentioned here and elsewhere. I met some as a youngster and others were colleagues later in life. It is indeed a pleasure to have them reappear in spirit as I write these words. To them, to the Wright brothers, and to the many whose names are not in the vocabulary of everyday life but who had a hand in helping to create the modern airliner and its engines, I dedicate what you are about to read and will hopefully enjoy.

The faces of many who played a role in this story are on the following pages. The list is by no means all-inclusive but highlights the people with the ideas.

Glossary of Abbreviations

AR	(Wing) Aspect Ratio
AEA	Aerial Experiment Association
BMW	Bayerische Motoren Werke
CAB	Civil Aeronautics Board
CEO	Chief Operating Officer
CFM	GE and SNECMA partnership
cu. in.	Cubic inches (61 cu. in. = 1 litre)
CW	Curtiss-Wright
ft	feet
GE	General Electric Company
HP	Horsepower
ICE or IC	Internal Combustion Engine
Ju	Junkers Flugzeug und Maschinenbau
Jumo	Junkers Motors
L/D	Lift to drag ratio
lb(s).	pound(s) force or weight
mph	(statute) miles per hour
NACA	National Advisory Committee on Aeronautics
NASA	National Aeronautics and Space Administration
OEW	Operating empty weight
PAA	Pan American Airways
Pan Am	Pan American World Airways
plc	Public Limited Company (British business law)
P&W	Pratt & Whitney
psi	pounds per square inch
RLM	Reich Luft Ministerium ([German] Air Ministry)
RPM	Revolutions per minute
RR	Rolls-Royce
SLS	Reference to 'Sea-level and Static' conditions
SNECMA	'Society for the Study and Construction of Aviation Motors' (French)

TBC	The Boeing Company
TOGW	Take-off gross weight
Toolco	Hughes Tool Company
TSFC	Thrust specific fuel consumption
TWA	Trans World Airlines (Transcontinental & Western Airlines)
UATC	United Aircraft and Transport Corporation
USAAC	US Army Air Corps
USAF	US Air Force
WW	World War (I or II)
XB	Experimental bomber
ZFW	Zero fuel weight

Dramatis Personae: Pictorial People Index

A photograph brings people described by names closer to life and to understanding the complicated things that men have built.

Thanks to the following institutions for providing the images in this section:

The Boeing Company: William Boeing, Donald Douglas, George Schairer, William Cook, and Joe Sutter.
General Electric: Sanford Moss and Gerhard Neumann.
The Igor I. Sikorsky Historical Archives
The author: Siegfried Decher and Anselm Franz
Pratt & Whitney Archives: Frederick B. Rentschler
GLMMAM Archive: Glenn L. Martin
Pan Am Historical Foundation: Juan Trippe
Wikipedia Commons: all others.

William E. Boeing
1881–1956

Adolf Busemann
1901–1986

William H. Cook
1913–2012

Glenn H. Curtiss
1878–1930

Siegfried H. Decher
1912–1980

Donald W. Douglas, Sr.
1892–1981

Anton H.G. Fokker
1890–1939

Anselm Franz
1900–1994

Howard R. Hughes
1905–1976

Hugo Junkers
1859–1935

Theodore von Karman
1881–1963

Charles L. Lawrance
1882–1950

Dramatis Personae: Pictorial People Index   xi

K.W. Otto Lilienthal
1848–1896

Charles A. Lindbergh
1902–1974

Ernst W.J.W. Mach
1838–1916

Glenn L. Martin
1886–1955

Sanford A. Moss
1872–1946

Gerhard Neumann
1917–1997

John K. (Jack) Northrop
1895–1981

Hans P. von Ohain
1911–1998

Frederick B. Rentschler
1887–1956

George Schairer
1913–2004

Thomas Selfridge
1882–1908

Igor I. Sikorsky
1889–1972

Joe Sutter
1921–2016

Juan Trippe
1899–1981

Frank Whittle
1907–1996

Orville Wright
1871–1948

Wilbur Wright
1867–1912

A note on the photographs in this book

The story about aeroplanes and their engines is best told with images to illustrate the artifacts discussed. Hence an attempt is made to bring the reader as close as possible to the text using photos and illustrations. Pictures of engines in museums are often mysterious, so their inclusion here is appropriate and hopefully interesting. Pictures of aeroplanes are everywhere, and in today's Internet age, such pictures are easily viewed there. It is nevertheless thought to be helpful to optimize the reading experience by having truly relevant pictures with detailed captions displayed within the text. But pictures of modern aeroplanes in service today are generally not included. It is my hope that pictures taken from the Internet are properly and honestly attributed to sources. If I have failed to do so, then please accept my apology and understand that my goal is to educate.

Acknowledgments

Putting what an engineer might think of as an interesting story into prose that a lay reader could enjoy is a challenge. This story cannot be written without description of the inventions and innovations of many. Further, it cannot be written without assistance from colleagues who have generously pointed the author in the direction of corrections and interesting and relevant information for addition to this story.

The role played by my father who steered me to follow his footsteps is difficult to describe or even to understand. It is ever-present nonetheless and to him I owe thanks for launching me into a most interesting and rewarding journey. The records he left behind were a rich trove of information of how we as a family went through the Second World War and how he managed a successful journey from the uncertainties of the war and the post-war period to a professionally rewarding career in the United States.

The institutional surroundings where one interacts or works have a profound effect on what one learns. People in them contributed substantially to this writing. Here I want to cite in particular The Boeing Company in the persons of Sarah Musi, Ben Linder, and Michael Lombardi, for their assistance in providing facts and images. Without the Museum of Flight in Seattle as a resource, this book could not have been written. In particular, Assistant Curator John Little was very helpful in directing me to interesting technical resources and to the Howard Hughes story specifically. In addition, discussions with docents Barry Latter and Jim Clyne yielded relevant historical and technical facts. For contributions from the General Electric Company, I thank Rick Kennedy for his gracious assistance with information and images.

Thanks are also due to Jon Proctor, the Igor I. Sikorsky Historical Archives, Mark Sullivan, Philip Edwards at the National Air and Space Museum, Douglas Culy of the Aircraft Engine Historical Society, Ronald Lindlauf, Engine Curator at the New England Air Museum, Nicholas Hurley, Curator at the New England Air Museum, and to Hans Toorens at the Historic Flight Foundation in Everett, Washington.

The images are credited in the captions. Many photographs are taken by the author at a variety of aviation museums. A very special thanks goes out to Miranda and Molly Blank whose skill with Photoshop helped clean up extraneous image aspects that might muddy the point being illustrated in the figures.

Finally, a heartfelt thanks is due to Mary, who showed immense understanding for the need to be immersed in this project.

Objectives

This is the story of the modern jet airliner and some of the innovators who developed it to the point where you and I can fly to an airport halfway around the world without it having to be refuelled. There is not much point in flying further. After all, the earth is a sphere!

As an aeroplane cannot fly without wings, it also cannot fly without propulsion. The capability of engines paced the development of aeroplanes. The histories of engines and of aeroplanes are interestingly intertwined. These histories include discoveries, an increasing understanding of the nature of air in motion, inventions, and the effort of thousands of engineers. These histories cannot be told without notice of world events, especially the two world wars that were the centrepieces of twentieth century history. All these aspects coalesced to allow the building of an ever-better aeroplane. The story is told without getting into the weeds of technical jargon and equations, but this cannot be entirely avoided as the subject is, at times, quantitative or technical in nature. It is hoped that the discussion motivates a future engineering student or evokes nostalgic appreciation by those who worked to bring about today's marvellous commercial aeroplanes. Even the casual reader may gain a better appreciation of the nature of the modern aeroplane that allows us to fly long distances in comfort and safety.

The history of airliners cannot be told without the history of engines. Neither can the story be told as a linear progression in time because it evolves on so many tracks: a number of engine manufacturers, airframe manufacturers, countries, and people. The availability of new engines allows for new ideas about aeroplane design and someone will invariably seize the opportunity to build a better aeroplane to further his position in the market place. The propulsion system, and it is a *system*, provides the force necessary to overcome drag at the speed we might wish to travel. It can be said that one of mankind's persistent wishes is to travel ever farther and faster, and propulsion always plays a leading role in determining an aeroplane's capabilities. While there are many technological components to a modern aeroplane, engine technology was an important, if not the central, dimension of the aeroplane. This history of propulsion in commercial aviation will hopefully make the centrality of engines evident.

The story told here does not evolve in a vacuum and there are events in the larger world and people whose work influences the story in significant ways. In addition, there are interesting facts and stories related to our tale but not central to it. These diversions are called out in footnotes so as not to interrupt the story.

The military dimension of aeroplanes is touched on only insofar as it relates to the evolution of aeroplanes of the type we call commercial transports.

The author resides in the American Pacific Northwest and has made a career in the proximity of The Boeing Company. That company has played a leading role in the evolution of airliners and, as such, will be cited possibly more often than a competitor might be. The reader will have to decide whether sufficient due credit is given to others. The reader may also have an opportunity to visit the museums that display so beautifully the work of the engineers and craftsmen who built aeroplanes. Among these are the National Air and Space Museum in Washington D.C., the Museum of Flight in Seattle, and the many others where some of the innovation details described here can be seen first-hand. I hope that reading this book may enhance the experience of the visitor to our aviation museums.

The colourful history of airlines as business enterprises in the US and around the world relates to this story only because some individuals in that industry were bold. Ralph Damon, the president of TWA (from 1949 to 1956), described air transportation as 'a race between technology and bankruptcy!' It certainly was, and is, for all concerned.

This text is not meant as a detailed story of the many projects that are cited. The stories of noteworthy aeroplane projects are described in the bibliography texts in much greater detail. The literature is rich with more aspects relevant to this story. This writing is a narrow overview of the relation between engines and the airliners they made possible. The timing of important steps is an integral element that hopefully paints a picture of how we travelled from aviation being a novelty to an everyday means of travel. The history of piston engines is particularly well described in two books that served this writer well. They are the volumes by Taylor and by Schlaifer that acquainted me with the technology developed a half-century before the jet age. With the development of the gas turbine, professional involvement with piston engine technology had passed to a lesser state of importance.

Much of the numerical information noted is drawn from the sources that made Wikipedia the storehouse of information that it is. While one should always be cautious about such data, a lifetime of working in the field allows judgment about the veracity of the information that is given and used here. Specifically, the power or thrust level of engines can vary significantly from a steady cruise rating, to a five-minute rating, or to war-emergency power as such a rating was once called. The quoted engine weights are similarly variable by inclusion or exclusion of fluids and other components. In the end, we can say that the level of precision needed for this story is not so severe that the possibility of misinformation will lead us very far astray.

The Internet is also a resource for videos that illustrate the sometimes difficult-to-describe gadgets and processes we deal with here. The reciprocating engines, in particular, are mechanically complex. The fundamentals of rotary engines, firing order in air-cooled radial engines, and sleeve valve mechanics, for example, are readily viewable for the reader with access to the Internet. Specific links to such resources are not given here as their permanence is uncertain and a search for such resources is rather straightforward.

The Propulsion Challenge

The relationship between aeroplane builders and power-plant builders has always been one where the airframers demand as much power as possible, with additional requirements for low engine weight, ease of maintenance, and minimal fuel usage. These parameters determine what the airliner can do physically and economically.

Power is delivered by a small number of heat engine types developed by inventors, engineers, and scientists. There aren't very many: the steam engine, the internal combustion engines invented by Nikolaus Otto[1] and Rudolf Diesel, and the gas turbine[2]. While there are many variations on these four engine types, there are no others that are practical. They are *heat engines* and the heat supplied is almost universally obtained from fossil fuels. These fuels are almost all burned with air. The air is everywhere one might wish to travel, thus it does not have to be transported. As a consequence, we have the interesting situation where, given reasonable efficiency, the amount of power produced is proportional to the amount of fuel and air that can be burned.

More air … more fuel … more power!

The central task of a heat engine is to produce a mass of gas – air or combustion gas – at high pressure. That is the requirement for driving a piston or producing a jet. The range of possibilities for producing compressed gas is very narrow:

---

1. In 1876, Nikolaus Otto, working with Gottlieb Daimler and Wilhelm Maybach patented the compressed gas, four cycle engine we know today.
2. In 1872, the American George Brayton invented the first commercial liquid-fuelled internal combustion engine. In academic circles, Brayton is honoured with his name on the thermodynamic cycle on which the gas turbine operates.

there are only two ways of doing it. The first is to confine the air to a cylinder and use a piston's mechanical power to compress it. There are various mechanical ways of doing this but they all amount to being the same process: displacement of the air in a confined space. That process is relatively easy and accounts for the reality that piston engines were the first of the gas-processing engines to be invented and perfected. The second way is to compress air dynamically through the forces inherent in aerodynamics. That is much harder and took much longer to achieve to a satisfactory degree, but achieve it we did and we are still working to improve it.

We should not forget that the very first engine to be invented to produce mechanical power was the steam engine. It benefitted from avoiding gas compression altogether and had only to raise the pressure of water, a significantly easier task. We owe a hearty thanks to the pioneers of that story, James Watt and the large number of tinkerers and scientists of the eighteenth and nineteenth centuries who set us on the road to wondering about ways we could develop other kinds of engines using air compression.

Since air compression is central to the idea of producing an engine and more air means more power, the question arises: how much air can an engine process? The amount depends on two parameters: a measure of the engine size and a measure of speed. For piston or 'reciprocating' engines, size is usually measured by the swept volume of the cylinders, so many cubic inches or litres. Speed, on the other hand, is most conveniently taken as the piston speed. In practice, the *mean piston speed*, defined as the stroke divided by the time associated with engine RPM (revolutions per minute), is limited by the ability of air and exhaust gas to flow past the necessary valves. Too high a mean piston speed leads to power losses. The typically accepted mean piston speed that allows for good flow rate and minimal losses is about 50 feet per second and varies for varying engine types, among them for ordinary automobiles, for racing cars, and for motorcycles. The practical speed depends on a number of design parameters such as the piston bore to stroke ratio (usually about 1), the valve design, the number of valves per cylinder and supercharging.

With piston speed as a constraint, the one significant way to increase the power output of an engine is to build it with larger displacement, the volume made up by bore and stroke. As we look over the evolution of reciprocating aircraft propulsion engines, the displacement grew over the years to the climax of the Pratt & Whitney R-4360, a behemoth of 4 rows of 7 radially arranged cylinders. The number 4360 refers to the displacement in cubic inches, which is equal to a displacement of 71.5 litres! This engine was typically rated for an output of

3,500 HP but could produce more under limited circumstances. The challenge of building this or any large engine rests in adequate cooling for the cylinders, keeping reliability high, and controlling maintenance cost in light of the very large number of moving parts that have to be lubricated and made to function within acceptable levels of stress.

The arrival of the gas turbine caused the discontinuation of further growth in reciprocating aero-engines. The relevant engine size is the frontal area and the speed is a significant fraction of the sound speed, which is about ten times larger than the 50 ft/sec limiting the piston engine. The gas turbine succeeded so well because it solved the airflow-supply limitation of piston engines. It did that by employing a very effective compressor that allows much greater power output merely by processing more air. Today, engines producing the equivalent of more than 50,000 horsepower are in service with modern airliners. A gas-turbine engine differs from the reciprocating engine and faces challenges that the latter's designer did not have to address.

The main feature of the gas turbine is that it has to deal with the continuous flow of very hot gases, hot enough to compromise most metals. These gas flows have to be handled in ways that lead to long engine life and minimize the possibility of surprise failures. The performance of the gas turbine is singularly dependent on the ability to achieve a high gas temperature at the combustor exit, *i.e.*, at the turbine inlet. The industry has achieved practical temperature levels with the development of good turbine-blade and disc-wheel cooling and the use of sophisticated materials for that engine component. Yet the amount of common petroleum-based fuel that is added to the compressed air is still short of what it could be for complete combustion of the oxygen in the air supplied. That possibility remains a promising carrot that may allow further improvement of today's good performance in the future.

In this book we will adhere to the use of descriptive quantities as traditional in the United States and in the aviation industry. For example, *feet* for altitude and the *statute mile* for distance even though the aviation standard is the *nautical mile*. Forces will be in *pounds-force*, mass in *pounds-mass* (*lbs*) (or tons, where a ton is 2,000 pounds-mass), temperature will be noted in degrees on the Fahrenheit scale and power in *horsepower* (*HP*). In Europe and elsewhere, the norm is to use measures in the metric system[3], but we will stick to American and aviation standards with an

---

3. The metric system units commonly used in this context are: kilowatts (kW, for power) and kiloNewtons (kN, for thrust). 1 Horsepower ≈ .75 kW; 1,000 lbs ≈ 4.45 kN.

ongoing apology for any mathematical stress presented to the reader. Conversion is easy enough, at least for the engineer.

The text includes discussion of some of the important details required to understand how a jet engine works. The description is without equations and mathematics but is hopefully sufficient for an appreciation of the technology and an understanding of the hurdles overcome by the pioneers central to that story.

> **'It seems very definite that the best aeroplane can only be designed around the best engine.'**
>
> Frederick B. Rentschler

Chapter 1

We are Flying!

Y ou are sitting in a comfortable seat on your way to a destination a thousand, perhaps ten thousand, miles away. On take-off, each engine of your airliner takes in and ejects about one *ton* of air *per second*, perhaps two! In flight, you cruise above most of earth's weather at 550 mph! You might argue about the seat comfort, but a nice person serves drinks and you have a chance to be with your thoughts, your own form of entertainment, or conversing in almost normal tones with your partner or someone new. You may even be enjoying the view of a beautiful world from your window seat. This is the mobility afforded by the modern aeroplane. Travelling was not always that easy: consider the travails of past travellers, explorers, and pioneers. The world has changed with modern air travel in many ways, not for everyone perhaps, but it has for those people active in the modern world's economy who can take advantage of it.

Fig. 1: You are here, going somewhere.

The development of flight as a means of transport is a fascinating story that spans a little more than a century. Most of the important technical steps that take us from air travel as a novelty to the state we enjoy today took place in three or four decades of innovation, roughly the years from 1935 to 1970. To be sure, innovation does not rest and evolutionary improvements continue to this day.

This is the story of the increments of understanding involved in enabling us to sit in that airliner. Many breakthrough ideas are involved. When and by whom were they implemented? Why were they important? And, finally, why were they critical?

Speed!

Speed of movement has always been an important driving force motivating innovation for military and commercial advantage. It was generally advantageous for armies to be in a new place fast, at least faster than an adversary. Over land, we witnessed the historical development of the use of horses and other animals, including camels and elephants, carts, Roman chariots, trucks, engine-powered tanks, and the like. Over water, ship design followed a similar evolution, and in the air, speed considerations were a consistent theme.

In the commercial world, history displays similar patterns for the distribution of goods, and the conveyance of people is consistently made better if it involves less travel time. Before the aeroplane, significant journeys over land or on the seas took weeks or even months, often with considerable shortages of comfort. The aeroplane shortened the time first to days, then to hours and generally improved the comfort level.

In addition to the transport of people and goods, speed was always important in the transmission of information. From the legendary Marathon runner to the Pony Express, carrier pigeons, and airmail, it was always desirable to have news as soon as possible. As long as a human being or a physical object has been involved as the messenger or message, they needed to be transported. Nowadays with the invention of radio and the establishment of modern digital infrastructure, news travels at the speed of light, limited only by delays involved in the generation and reception of the message.

Thus, while information may not need an aeroplane, speed as a parameter describing flight is singularly important. The speeds quoted for airliners in this book will be *cruising speed*. Generally, that is the speed that leads to good fuel consumption performance. After all, range was always important and fuel wasted by going fast was not in the interest of an airline. Typically, aircraft have maximum speeds that are greater than cruising speeds, but maxima tend to be primarily of interest for military combat performance purposes, and to indicate to the airline pilot: do not exceed this speed!

The air in which we fly

Since flight is in the Earth's atmosphere, one cannot talk about the speed of aeroplanes without recognizing the limitations the atmosphere imposes. The atmosphere is a domain of varying air pressures and temperatures at various altitudes. For purposes of flight, the atmosphere consists of a lower layer where weather is active and variable, and a higher layer where the weather is less variable,

Fig. 2: The weather and visibility are better up here, but storm turbulence can easily penetrate this calm world.

because it is less affected by the influence of solar energy input. These layers are the troposphere and the stratosphere. The boundary between them, the so-called tropopause, lies as low as 20,000 feet in polar regions, and may approach 60,000 feet in the tropics. For a so-called standard atmosphere, the tropopause is fixed at 36,090 feet.

Every aircraft has a 'service ceiling'. This is the maximum practical altitude at which a particular aeroplane can fly. The aeroplanes of the 1930s had service ceilings of about 25,000 ft, and were unpressurized. But nowadays an aeroplane would never be unpressurized in service at that altitude. The passengers would surely object ... if they made it to their destinations alive! The human body can tolerate the lower pressure at higher altitudes but is subject to a number of deleterious issues. These are numerous and complex; suffice to say that, if cabin air pressure or flight altitude is held to a level under 5,000 ft, the average passenger should be able to travel without ill effects. At that altitude, the normal temperature is some 27 degrees Fahrenheit lower than it is at sea-level. Heat must be provided or warm clothing worn by the passengers. In the early days of airline aviation, heights of 8,000 ft were routinely reached to cross the Rocky Mountains. Airlines provided oxygen for passengers and crews as needed.

Flight at lower altitudes is always subject to some influence by the weather. In the stratosphere, above the weather zone, the air is much less likely to be turbulent. Air pressure at 35,000 feet (near where modern airliners cruise) is one quarter of sea-level pressure. There are good reasons for flying high. Another is to avoid encountering mountains!

The Speed of Sound

In any medium, including the air, there is a fundamental speed at which pressure waves propagate. This is the speed with which information, pressure waves if you will, is passed on within the air to tell it to get out of the way of an approaching wing or propeller blade. Visually they may be compared to the waves that emanate from the bow of a ship. Very small pressure fluctuations are perceived as sound waves; hence we describe the air's characteristic wave-propagation speed as a *sound speed*. The limited sound speed is one feature of Earth's atmosphere that shapes the manner in which flight within it can take place.

Objects moving at speeds close to the speed of sound generate stronger waves. In turn, these waves have strong effects on the drag experienced by the moving object. This is true even for the propeller blades that powered all aeroplanes in the early years of aviation and limits the speed of today's airliners to less than the sound speed.

When strong enough, the waves created by a moving object are *shock waves*. Air crossing a shock wave experiences a pressure increase. Shock waves affect the flow far from the body. That means that a lot of power is invested in creating and maintaining them. They also have a profound effect on the pressure over the wing or propeller blade. Briefly, they can cause the flow to fail to conform to the shape of the wing or blade and that results in a thick wake[1] that should ideally be as thin as possible. The drag increase associated with increasing speed is commonly referred to as the 'sound barrier'. This is not really a barrier but the reality that the power available may simply be insufficient to move the wing or propeller blade any faster.

The speed of sound in air depends solely on its temperature. It will therefore vary with altitude. At sea-level, on a so-called 'standard' day, the sound speed is about 760 miles per hour, and in the colder stratosphere it is a more uniform 660 mph. These speeds are good reference points for judging speed performance of aeroplanes in various periods. In practice today, speeds near that of sound are quoted as a fraction of the sound speed. This fraction is the Mach[2] number, where movement at *Mach one (M = 1.0)* is at the speed of sound. The use of the Mach number frees the discussion from consideration of temperature conditions at the time and place an aeroplane is flying.

The Airline Business

The business of building and operating airliners is a technical and an economic challenge. Both aspects have quantitative dimensions that have to be considered. Numbers are critical and we have to take an interest in how they are determined. Airlines have to make a profit.

Running any business invariably involves keeping track of numbers. Some of the numbers cannot necessarily be controlled or accurately determined, so one has to employ estimates. As a result, proceeding into the future carries some risk. Fuel price is one dimension of the airline industry that has shaken it more than once. Further, the industry buys airliners with a period of years between order and delivery, and that new airliner should function economically for decades after it goes into service.

---

1. The wake is the air that has been accelerated by friction with the aeroplane and tends to be dragged behind it. It is similar to the wake behind a moving boat. It can be thought of as a 'negative jet'.
2. After Ernst Mach (1838–1916), Austrian physicist and philosopher.

For example, an airliner might be designed to last thirty years. Cracks associated with metal fatigue of the structure are normally the critical life-determining issue. Take-off/landing events, together with violent weather encounters, tend to be the hardest loads on the airframe. Today, a long-range aeroplane might be designed for 40,000 cycles while a short-haul aircraft might be designed for 100,000 cycles. In either case, it flies for 10 to 20 minutes of every hour of its existence. That suggests a heavy time utilization of the aeroplane, and loading and unloading must also be taken into consideration.

If you think concerns for profitability over these long periods are difficult to manage, try to find answers to the questions that the airliner-maker will have to address *before* he launches a new aeroplane type. His thinking has to involve time periods even longer because the production run of the aeroplane will, hopefully, be many years long. He has to make sure that the enormous investment is repaid, with interest, and a profit made. If you glean that this industry is subject to economic turbulence, requires people with a high degree of risk tolerance and financial resources to back it, then you have a good idea of the environment in which this story is set.

A quick look on the Internet for 'defunct airlines' yields a rich history of airline names that are no longer. Some vanished airlines have been acquired by the large carriers that fly in service today. Others never did manage economic operation in the long run. In a curious way, many of the names that disappeared carried a regional aura: Western, Eastern, Southern, Northwest, Pacific, Mohawk, etc, while others reached across the US with their names: Continental and USAir, among them. Some even had a global aspirational reach: Transworld and Pan American. Finally, some names did not include geographical references and were associated with an individual owner. Hughes AirWest and Braniff come to mind. Only the older reader will be transported to the time when these airlines operated. They are now part of history.

Hindsight is an advantage that allows the historian to focus on the development avenues that bore the most fruit. Thus we can look back and concentrate on the elements that led to the present state of the airliner. As far as the technological issues are concerned, the principal actors are the aeroplane-manufacturers and the engine-manufacturers. The latter are included because it has always been true that propulsion plays the greatest role in determining an aeroplane's design. It is therefore critical to include the development of engine technology because it leads the parallel evolutionary paths of aeroplane-engines and aeroplanes. There were, of course, other technical fields whose inventions and understanding are critical to the successful operation of today's commercial aviation.

The Aircraft Engine Builders

The history of engines is complicated by the ever-changing business arrangements of the people involved. Many attempts have been made to establish successful enterprises around engine-manufacturing, and in the US these culminated in companies like Curtiss-Wright and Pratt & Whitney playing dominant roles in the technological development of reciprocating aircraft-engines. There were and are others, of course, Allison chief among them, but Curtiss-Wright and Pratt & Whitney were central to the large aeroplanes we will consider here. The history covered includes the Second World War in important ways. The military dimensions are relevant because the war effort underlies a lot of the motivation of engine development during the war. In the military, bombers and transport aircraft were demanding high performance and the associated technology is naturally passed on to civil airliners, later perhaps, but surely.

To be semantically correct we might note that a piston-engine driving a propeller is, strictly speaking, a 'jet' engine. The propeller's function is to create a jet. To be clear with our descriptions however, we will adopt the normal convention and limit the label *jet engine* to the new-technology engine developed with continuous combustion and refer to the internal-combustion engines driving a propeller as piston or reciprocating engines.

During the war and the post-war years, the history of jet-engine development in the United States was dominated by two companies: General Electric (GE) and Pratt & Whitney. Curtiss-Wright was not successful in adapting to the Jet Age, at least in producing jet engines. The company made an attempt by licensing the British Armstrong-Siddeley *Sapphire* engine in 1950. It developed the engine and sold it to the US military as the J65 for a number of military applications. Allison and Westinghouse did, for a time, go on to be players in the jet engine industry. There were and are other jet-engine manufacturers who worked successfully in aviation, but the airliners of today employ primarily, though not exclusively, the large American engines made by GE and Pratt & Whitney. On the world stage, on which the airline industry performs, the British firm Rolls-Royce, the French firm SNECMA (now Safran) and others also have a strong hand in modern jet engines for airliners. The other manufacturers are most commonly consortia of industrial concerns outside the US that often include one of the 'big three', General Electric, Pratt & Whitney, or Rolls-Royce. GE and Pratt & Whitney are private companies. Rolls-Royce was, for a time, a public company partially owned by the British government and is now again in private hands. The reason is interesting and will be mentioned when the engines are discussed later in our story.

When describing engines, it is customary in the US to talk about *horsepower* (which we will henceforth abbreviate as HP) for piston engines and *pounds of thrust* for jet engines. In both cases, the rating is stated in terms of performance at sea-level conditions in a stationary situation. This is referred to '*sea-level, static*' (SLS) conditions. Both the power and the thrust of engines are relatively easy to measure there. Performance levels decrease as an aeroplane flies at higher altitudes because the air density is lower than it is at sea level, as are pressure and temperature. For piston engines, this tendency towards reduced performance at higher altitude is mitigated somewhat by the use of superchargers (mechanically driven compressors to raise the density of the engine intake air) or turbochargers (same kind of compressor, but driven by the power in the hot engine-exhaust gases). Fortunately, the need for power or thrust also decreases with altitude. For jet engines, thrust decreases with altitude roughly in proportion to the decreasing atmospheric pressure.

The Airframe Building Companies

The domain of the civil airliner, like the engine business, includes many tries at long-term business success. We know who the dominant American airliner-builder is today, The Boeing Company. Today's company has roots that extend to other airframe-builders with whom it has joined forces. All other companies active in the industry build primarily for the military today, but their contributions are significant. Major competitors travelled a long way along the airliner-manufacturing road and chief among them are the Douglas Aircraft Company and the Lockheed Corporation. There were others as well, the Glenn L. Martin Company, Convair,[3] among them. To describe the road to the present day, the focus will be on the three majors of that history (Boeing, Douglas, and Lockheed) by citing the airliners they produced and how they fit into the longer story. Paralleling the American story are foreign manufacturers. Some were and are indeed significant pioneers.

Performance numbers will be quoted that are representative of the aeroplanes called airliners. These numbers are not meant to be exact. They can be gleaned from many sources that may even provide conflicting values. The goal is to describe the aeroplanes in general terms, even as we understand that any aeroplane type can have variants as it is being produced. It can also be designed or operated to fly various missions that might involve a longer flight distance with a smaller passenger or

---

3. A merger of Consolidated and Vultee Aircraft Companies in 1943.

freight load. It can even be built with engines from different manufacturers. While we may be dealing with less-than-exact numbers, the aeroplane-builder and the airline do insist on precision. Most of the numbers describing performance with a high level of certainty tend to be proprietary and not available to the general public. So, we do the best we can with what we can find!

The best way to describe the characteristics of airliners is to use the performance given for the early models, because it is these that initially convinced airlines to buy them. The subsequent models of these early offerings were often and significantly improved, especially when the production runs were long. The later offerings were not necessarily for improvements as such, but were often models designed to meet specific market niches. This tendency makes it difficult to talk definitively about the design and performance of an aeroplane type and the reader should view cited aeroplane characteristics in that light.

Chapter 2

Aviation is All the Rage

Perhaps the idea of conveying passengers by air to distant places germinated with Rufus Porter. Porter was an inventor and publisher in the mid-nineteenth century. When the fever of the Gold Rush in California was at its hottest, he proposed to take a large group of fortune-seekers to California by air! He published a work titled *Aerial Navigation: The Practicability Of Traveling Pleasantly And Safely From New York To California In Three Days, Fully Demonstrated*. He proposed to travel at speeds of 100 miles per hour (a remarkable claim at that time) in a dirigible powered by a steam engine turning a wheel resembling a modern propeller. The large, long airship was to have a 'salon' suspended from it with a large number of windows and an American flag fluttering at the stern! Three meals a day were to be served to the travellers as they crossed, not from the banks of the Missouri, but from New York, to seek their fortunes in the hills of California. A number of investors and potential customers enthusiastically supported the scheme, but, as a modern person with the hindsight of history can easily imagine, the scheme came to naught and the invested funds evaporated. The hopeful and disappointed customers had to find a less-attractive alternative to reach California than in what Porter called the *Air Line*.

Air travel would have to await the development of propulsion technology that was up to the job! A switch in thinking from *aerostatics* to *aerodynamics* to produce lift would also have to take place. Nevertheless, Rufus Porter will be remembered for his many other inventions, most now viewable only in museums, and for *Scientific American* magazine that has weathered the storms of time.

The credit for first successfully flying the airship conceived by Porter is given to Jules Henri Giffard, a Frenchman who built and flew one in September of 1852. Giffard's propulsion consisted of a three-bladed, paddle-like propeller, driven by the only available engine at that time, a steam engine. The idea did not go much further until much later. The 3½ horsepower steam engine and the technology was estimated to weigh in at about 120 lbs/HP.[1]

---

1. Engine weight in *Pounds per Horsepower* (lbs/HP) as an engine performance measure is singularly important and will be cited often as the history of engines unfolds. In the jet age, the thrust produced per pound of engine weight (Engine thrust to weight ratio) will be similarly central.

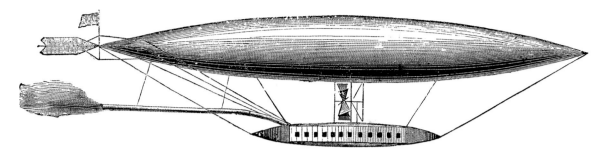

Fig. 3: Under the banner 'Best Way to California', the proposed Porter airship powered by a steam engine! (istockphoto.com)

In the mid-1890s, two serious attempts at powered dirigible flight were undertaken. One was by Sir Hiram Maxim (1894) and the other by then Secretary of the Smithsonian Institution, Dr Samuel P. Langley (1896). The steam engines employed had improved to a weight of 5–7 lbs/HP. Steam engine technology was heavy for flight. Perhaps the gasoline engine would do better.

In 1898, the aviation pioneer Alberto Santos Dumont flew a gasoline engine powered dirigible with an engine weight of 19 lbs/HP. That was no improvement over the steam engine but he bettered it with an 8.8 lbs/HP Clement engine in 1903. That was still heavy, but it was starting to look like engines light enough to fly could be built. To put the year 1903 in some perspective, we note that it was the year that an automobile made the first trip across the continental US. The 4,500 mile journey took 63 days!

In the meantime, curious men, and they were all men, looked to the birds for inspiration and began to try to understand the nature of aerodynamic lift with gliders. One genius, far ahead of his time, was Leonardo Da Vinci (1452–1519) who seriously examined the mechanics of flight and the possibility of imitating birds, and designed a rig that might allow a man to fly. To the best of our knowledge, it was never built in a serious attempt at flight. Even if it had been, it would not have succeeded. In the late nineteenth and early twentieth centuries, however, other people did try to fly novel and unsuccessful contraptions that provide humour to today's viewers of video clips. There were, however, serious people trying to understand flight. Otto Lilienthal, the Wright brothers, Octave Chanute, and others comer to mind.

Aerodynamics Pioneers

After persistent and disciplined research, the Wright brothers, Wilbur (1867–1912) and Orville (1871–1948), answered a question people had posed since the time of Icarus, and the answer is: a man can indeed devise a vehicle to allow him to fly like a bird. What they accomplished is an engineering masterpiece of understanding details, adapting what they knew to new circumstances, and making quantitative determinations of information that had not previously been known. Their story is well told elsewhere. It climaxed in December of 1903 with their first powered, sustained, and controlled flights in the coastal dunes of North Carolina. Their accomplishment has many dimensions. These include the notion of using wind tunnel tests. The results led to an understanding of the performance of aerofoil shape and realistic wings as lifting devices and the drag penalty that these wings exact. The Wrights, with the assistance of their mechanic Charles Taylor, developed a relatively light-weight gasoline engine to deliver the power required. Their most significant contribution will remain the satisfactory means of controlling the flight so that it is safe. Not to be overlooked is that they, and others elsewhere and earlier, succeeded in building the wings to be sufficiently light.

The engine used by the Wright brothers for their first flight may be used as a reference to measure the performance improvements made by engine builders since then. These improvements were gigantic to be sure, because the start was modest. The Wright Flyer engine was a 4-cylinder, inline design that had

Fig. 4: A Wright flyer at rest showing the control surfaces in front of the pilot and an indication of the wing twist to control the roll of the aeroplane. Note the two chain-driven counter-rotating propellers and their drive shafts.

a displacement[2] of 201 cubic inches or 3.3 litres, about the size of a modern automobile engine. It weighed 180 pounds and produced 12 horsepower. Consider the performance contrast to the modern automobile engine that, for the same displacement can provide over 200 horsepower and weigh much less! The Wright engine weighed 15 lbs/HP. That indicator of technological performance was significantly less than what the Langley-Manly engine had achieved with an amazing 2.58 lbs/HP. The Wright engine, however, was an independent achievement and one that was first to fly!

These numbers might be compared to what a biological system can do. For example, a horse, weighing about 1,000 lbs might produce *one* horsepower! A human, like the athlete Bryan Allen who powered a flight across the English Channel in the Gossamer Albatross[3] in 1979, might better this number, but not by much and certainly not for long!

Each cubic inch of Wright engine displacement produced 0.06 horsepower. These numbers are modest and the brothers had already included the innovation of a cast aluminum crankcase in their engine to reduce the engine weight. In 1910, the brothers improved the 1903 engine to the point where it produced 30 HP, more than doubling its performance to 6 lbs/HP. Looking into the future, as the era of piston engines for aeroplanes approached its end, improvements from many quarters were to eventually raise performance to about 1 lb/HP and 1.0 HP per cubic inch.

At the New England Air Museum is a display thought to be the oldest surviving Wright aero engine in running condition (see Fig. 5). It dates to about 1910. It is reported to produce between 28 and 42 HP, has a displacement of 240 cubic inches and weighs between 160 and 180 lbs. Note the flywheel on the right that the builders of rotary engines thought to eliminate by having the cylinder array rotate.

Two aspects of the early Wright Flyer engine are illustrated in Fig. 4. The first to note is the very small fuel tank located just below the upper wing to the left of the prone pilot. The fuel supply was not meant to last for long. Note also the absence of a radiator for the water-cooled engine. The water coolant is maintained in the engine and allowed to heat up during the short flight. The long, black,

---

2. The displacement is the volume swept by all pistons within the cylinders of a reciprocating internal combustion engine.
3. A pedal powered aeroplane with a wing aspect ratio of 19.5, weighing 70 lbs empty, and 215 lbs with its human power plant. It is estimated to have required 0.4 HP continuous power output for the 22 mile crossing of the English Channel. Allen did it for 2 hours and 49 minutes!

Fig. 5: A circa 1910 Wright water-cooled aero engine at the New England Air Museum. The 240 cubic inch engine is estimated to produce 28–42 HP. The hose connections at left are for coolant from the pump (lower left) to the jackets around the cylinders and ultimately to a radiator. Note the flywheel at right. (Courtesy New England Air Museum).

vertical tank to the right of the pilot maintains the coolant water pressure. The engine is designed to run at maximum power (there was no throttle and no carburettor) for just a few minutes! Ignition was provided by means of spark plugs with cam-driven interrupters metering the electrical current from a twenty pound (!) magneto. Considering what we can do today, these twenty pounds were quite a weight penalty. Their use of an aluminum crankcase, instead of the then common cast iron, helped reduce the engine weight.

The brothers tried very hard to protect the intellectual property they had amassed as they continued development of their flying machines, as much out of the public's eye as was practical. In the years after the first flight, the public was not well aware of their accomplishment, but other flight enthusiasts were working on controlled flight in a kind of 'quiet' race. Among the others were individuals in Europe, in North America, and Alexander Graham Bell in Canada. Bell was experimenting with kites and gliders in that period and knew of the work of his friend Samuel Langley. Bell's gliders were novel but quite 'draggy' and heavy. When he learned of the Wright brothers' flight in 1905, Bell assembled a handful of young people to help him with building a practical aeroplane of their own, what Langley called an 'aerodrome'. Two of these individuals were Glenn H. Curtiss, a successful builder of light-weight engines for motorcycles and Thomas Selfridge. Selfridge had been appointed to the new Aeronautical Division of the US Army Signal Corps and worked with Bell on an informal basis. Bell's group called itself the 'Aerial Experiment Association' when it was formed in 1907.

In the meantime, the Wright brothers were eager to sell the aeroplane idea to the US Army and had to protect their ideas as they were going to be more exposed to the public eye when the US government was involved. To help protect themselves, the Wrights turned to the US Patent Office where, on 22 May 1906, they secured a patent[4] that was to protect and financially reward them for seventeen years from infringement by other aeroplane-builders. That was the hope but, unfortunately, it was not realized.

Bell's AEA was not to play a lasting role in aeroplane design. A number of increasingly successful aeroplanes were built, the last of which were the *June Bug* and the *Silver Dart*. These aeroplanes were quite similar to the Wright flyers but also differed in innovative ways. The similarities included biplane wings, vertical tail in the rear and horizontal control surface in front. The innovations in the *June Bug* included an air-cooled Curtiss engine, tricycle landing gear wheels, and ailerons as we know them today in lieu of the Wrights' twisted wing tip. The *June Bug* set a one-kilometre flight length record on 4 July 1908 with the goal of winning the *Scientific American Prize* for that achievement. After difficulties with overheating, the later and improved *Silver Dart* was fitted with a water-cooled engine. The history of air- versus water-cooling had begun.

During this time, the US Army recognized the military value of the aeroplane. In a little more than a decade after the first flight, the world was to be at war and that was to provide an impetus to develop the aeroplane as a weapon, specifically to find out what the enemy was up to. Of course, there was also the need to keep the enemy from doing the same things! That motivation was not yet well defined nor fully appreciated, but the aeroplane ultimately became an important, though not necessarily a decisive, part of the First World War. The Wrights were aware of the aeroplane's potential in war and tried to obtain aeroplane production contracts that would carry their business interests far into the future.

---

4. US Patent number 821,393 for a 'Flying Machine'.

The Wrights began a campaign to interest the US Army in the military value of aeroplanes. They viewed the military as the best and most rewarding customer for aircraft. At a demonstration of their aeroplanes in March 1908, flight with two people was to be shown as practical. To the Wrights' discomfort, because they knew of the Bell and AEA connection, Thomas Selfridge was the army official evaluating the Wright aeroplane. The Wrights were concerned about their technology finding its way into a competitor's aeroplane. The need to sell aeroplanes to the army, however, trumped concerns and a flight piloted by Orville Wright with Selfridge as an observer was arranged. Unfortunately the flight ended in a crash and Selfridge[5] was killed. He was the first fatality of powered flying. Orville himself was seriously, though not fatally, injured.

The successful building and flying of the Silver Dart brought the AEA to its goal. That and the death of Thomas Selfridge ended up as reasons for the dissolution of the AEA in March 1909, after eighteen months of operation. Of the people involved in AEA, Glenn Curtiss carried on his aeronautical interests in ways that we will encounter again in our tale.

The Wrights claimed that the idea of a 'flying machine' was in the heart of their patent and thus any other flying machine built should earn them royalties. That idea was vigorously fought in the courts by some and largely ignored by others. During the resulting disputes, Orville Wright retired in 1915 and the matter was carried forward by a partnership between Glenn L. Martin and the Wright interests. Under these circumstances, the aileron as an innovation by Curtiss and the AEA was an important component in the legal dispute between the Wright-Martin Company and Curtiss. It was resolved by pressure from the US government because the US Army needed aeroplanes when the US entered the war in 1917. The legal fight had hampered US aeroplane production for the war. After the war, interest in continued litigation faded and there the matter rested.

The Wright-Martin Company was a short-lived partnership that Glenn L. Martin left to focus on his own ideas regarding aircraft manufacturing. History will record that he succeeded well in that enterprise.

Irrespective of the legalities of aeroplane designs, many an aeroplane builder tried to emulate what the Wright brothers and others had so well demonstrated. Many did it successfully. In the decade before the war, the improving capabilities of aeroplanes were shown off to wildly enthusiastic crowds who came to witness

---

5. Another noteworthy fatality in aviation was that of Otto Lilienthal who investigated controlled flight with gliders. One such flight cost him his life in 1896.

the feats of magnificent men and their flying machines. As these machines became more capable, passengers were taken for demonstration rides.

While the year 1903 is noted in aviation history as the date for first powered and controlled flight, the year 1909 also merits mention for a number of reasons. These include achievement of over one hour of flight by a number of aviators and new speed and distance records. A wide variety of engine types were tried and flown. These include air- and water-cooled engines in in-line, Vee, opposed, as well as radial configurations. Among these is the famous French Gnôme rotary engine. Finally, the year also saw the introduction of mechanically actuated inlet valves that had heretofore been opened by pressure differences, *i.e.*, by suction on the piston-cylinder volume.

Aviation Innovators and Promoters

History reveals that daredevil flight demonstrators necessarily follow inventors to take technology to the next step. Its novelty and potential had to be shared with the public so that money and reputations could be made. Public exhibition was an aspect of the work the Wright brothers did not cherish. They worked in secret and did not call attention to their activities. They sought a high level of perfection for their 'flying machines'; the world is not, however, patient. After their successful experiments, it was time for new ingenious people step onto the world stage. Their stories are many.

One such concerns Glenn H. Curtiss. In the few years he was associated with the Aerial Experiment Association, and thereafter, he brought the aeroplane's capability to the attention of the public with performance records and award of prizes. In 1910 he formed the Curtiss Exhibition Team that performed flight events that were widely promoted for viewing by the public. He was a good promoter as well as innovator.

An important dimension of his work was his pioneering breakthroughs in naval aviation, a field for which he may be considered an important contributor. He devised a practical pontoon configuration that allowed an aeroplane to take off and land on water. He and a member of the team, Eugene Ely, also helped demonstrate the practicality of landing an aeroplane on the deck of a ship in 1910 and 1911. Sadly, in that same year, Ely fell victim to the risks inherent in exhibition flying with a fatal crash. For the later war effort, Curtiss designed the JN-4 (Jenny) and a large (for its time) amphibious three- (later, four-) engine flying boat, the Curtiss NC.

The year 1911 was also the time when an important innovator in Tsarist Russia named Igor I. Sikorsky set his mind to building (among many other designs) an aeroplane large enough to become the first what we might call an airliner today. It was a large biplane with four in-line water cooled engines of 100 HP each. It was called the Russky Vitiaz (Russian Knight) built by the Russian Baltic Railroad Car Works. For its designer it was the S-21. Its looks did indeed betray a railroad car builder's influence! It flew in 1913. The cabin had large windows, and a 'porch' was installed at the nose of the aircraft. In similar military aeroplanes built for the later war, this kind of porch often allowed a machine gun to be installed.

After trials with several engine configurations, settling finally on four tractor (pulling propeller) engines, it was established to be able to carry seven passengers with a crew of three. Its performance can be described as modest: speed 56 mph, service ceiling 2,000 feet, and a range of 106 miles. It was a start, nevertheless. While this airplane never saw service (due to an accident involving a dropped engine from another aeroplane flying overhead!) it can claim an early step into the world of transport of people by air. In this story, we will encounter more contributions by this inventive genius after he left post revolution Russia.

Back on American shores and after the war, Orville Wright and Glenn Curtiss, who had retired in 1915 and 1920 respectively, stepped back from active involvement in aviation. The companies that followed continued with construction of aircraft and aircraft engines. Wright Aeronautical did not succeed in making a mark with construction of aeroplanes but did so with engines. Curtiss succeeded with both aeroplanes and engines. History reveals that their focus on engines carried them the furthest and, in time, the later companies had to join forces to survive. Thus Curtiss-Wright proceeds into the future. Interestingly, the engines the company produced would predominantly carry the name 'Wright'. The irony of that merger stems from Wright's inept management that resulted in the creation of Pratt & Whitney as a very competent competitor.

The years before and after the First World War saw a number of other significant corporate creations on the heels of aviation arising as more than a novelty. Their influence would persist for many decades. The Lockheed Corporation (which was known before 1926 as the Loughead Aircraft Manufacturing Company) and the Glenn L. Martin Company were founded in 1912. William E. Boeing started the Boeing Aeroplane Company in 1916.[6] In 1921, Donald W. Douglas founded the Douglas Aircraft Company. These companies were to take full advantage of the engines brought out by the engine industry to develop the airliners of the day.

---

6. The company created in 1916 was the Pacific Aero Products Co. and the name was changed in 1917.

Parenthetically, the well-known aeroplane designer and builder John (Jack) Northrop started his career as a draftsman at Loughead. In 1923, he went on to Douglas, where he had a hand in the design of the Douglas Round-the-World Cruiser and helped Jack Ryan with the design of Charles Lindbergh's *Spirit of St. Louis*. In 1927 he returned to the renamed Lockheed Corporation where he worked on the civilian transport called the Vega.[7] He founded the Northrop Corporation in 1932 and started a long line of successful and interesting aeroplanes. Northrop did not start a new aeroplane company in the war period, nor did his future company build airliners, but his legacy influenced many of the steps taken by the aviation industry.

As the war conflict ended, the need for military aircraft was reduced dramatically and these became surplus, available for the public. After all, this was the war to end all wars! The ex-military aeroplanes were a boon to the public during the early interbellum period. This was especially true among the victorious Western Allies in whose countries barnstorming, air races, and similar feats were popular entertainment. Defeated Germany did not participate fully in this technological party because it was forbidden to build military aeroplanes by the war-ending treaty.[8] However, since motorless glider aircraft were not covered by the treaty, interest in gliding thrived in Germany[9] as did interest in fundamental research into the physics of fluid motion, specifically air.

As peace returned to Europe, the notion of commercial aviation to carry freight and passengers dawned in Holland. The airline KLM was first to go into operation. The company was created in 1919 and started operations in the following year. By 1926 it offered flights to a multitude of cities in northern Europe, using primarily aircraft made by the Dutch manufacturer Fokker. It still operates today and thus is the longest-lived airline with its original name. Its operations were interrupted only by war and occupation in the 1940s.

Hugo Junkers

In the history of aeroplanes and aeroplane engines, there is one German pioneer whose influence is important from technical and industrial viewpoints. Hugo Junkers (1859–1935) formed an aircraft building concern and a number of

---

7. An aeroplane used widely, including by Amelia Earhart.
8. The 1919 Treaty of Versailles' Article 198 prohibited Germany from having an air force and from producing anything related to military aviation.
9. See *Akaflieg*, an academic flying club with wide membership, esp. during the 1920s.

innovations are to his credit. His aviation interests date to the World War I period and extend to his death in 1935. His work centred on the three dimensions of aviation: aeroplane construction, engines, and operation of air transport services within the limitations imposed by the Treaty of Versailles. Junkers nevertheless saw a future in commercial aviation. The culmination of his aeroplane building effort was the realization of the Ju52/3m, a successful airliner (a low-wing trimotor, hence 3m) described later in our story. His commercial transport business interest was motivated by the need to sell aeroplanes to air transport services. To that end, he had a strong hand in the creation (in 1921) of an airline that ultimately became Deutsche Lufthansa. It prospered before and after the Second World War to the present day.

Good aeroplanes needed good design ideas. His included an early use of all-metal airframe construction (1915). The use of corrugated sheet iron (yes, iron!) is associated with Junkers as is the later and widely used corrugated aluminum that is so characteristic of the aeroplanes of that time. He pioneered low wing monoplane design ideas that were widely adopted by others.

Good aeroplanes also needed good engines and his prewar involvement in engines served him well in that his firm, Junkers Flugzeugbau and Motorenbau, incorporated construction of both engines and aircraft. The company went on to build a series of liquid-cooled aircraft engines whose performance improved steadily over time.

The difficult economic times after 1929 led to Junkers' sale of other assets to preserve his aviation company. The company struggled. It did not help that Junkers was not enamoured with the new Nazi government when it came to power (1933), neither was it with him. After his death in 1935 at the age of 76, the government took over his aviation assets and turned the capabilities into a part of the German war machine. As Hugo Junkers was a pacifist, this turn is particularly ironic. The Second World War saw the use of a number of Junkers aeroplanes.

American Enthusiasm for Aviation

In the United States, the use of aeroplanes for commercial purposes became more serious not very long after the Wright brothers and others showed what aeroplanes could do. The most significant commercial endeavour was the transport of mail that was inaugurated on 15 May 1918 after earlier trials dating back to a time before the war. The post-war period was a time for setting records of speed, altitude, and distance, constantly pushing the capability of aeroplanes past the achievement boundaries of the day. There were fame and prizes to be won for new accomplishments. One noteworthy challenge was the successful crossing of the Atlantic, from New York to Paris, in 1927 by Charles Lindbergh. His accomplishment was directly linked to the training and experience gained in the early 1920s as one of the American airmail flyers.

The non-stop flight in a single engine aeroplane set the stage for accomplishing that feat with larger aeroplanes actually carrying something more than just the pilot. It did not take long: in 1928, Amelia Earhart crossed the Atlantic as a passenger in Fokker F.VII that had a two-man crew and crossed it again solo in 1932.

The aviation spirit struck those who understood that the future of aviation was forged by young engineers trained in the science and the technology of flight. Notable was the contribution of Daniel Guggenheim and his son Harry, who established a fund for the 'promotion of aeronautics' on 16 June 1926. They donated $3 million between then and 1930 to establish university-level education programmes in the US. The gifts were made to many of the most important American engineering schools and universities of the day. Most of these programmes, with their associated Guggenheim buildings, remain active today.

Engines in the Early Twentieth Century

Before the First World War, most serious aircraft engine-development effort was carried out in Europe. Great strides were made in France, Germany, and Great Britain after about 1910. Engine manufacturers were numerous and some went on to be viable for a long time thereafter. An early French builder of renown was Hispano-Suiza. Their innovative engines were used in large numbers of First World War Allied military airplanes. They were water-cooled V-8s built with an all-aluminum body except for cylinder barrels and moving parts. A post-war example is shown in Fig. 7. The Wright-Martin Company was created in 1916 to produce the Hispano-Suiza engines under licence paralleling other arrangements to produce European water-cooled engines in the US.

American engines were also built, notably by Glenn Curtiss, formerly with the AEA. The Curtiss Aeroplane and Motor Company, founded in 1916, built a number of engine types with the OX (90 HP) and OXX (100 HP) engines being most successful. These engines featured a clever valve actuation mechanism where a single pushrod moved both intake and exhaust valves at the top of each cylinder. Curtiss engines were purchased by the British military. In all, over 12,000 OX-5 engines were built for many types of aircraft including the Curtiss JN-4. Production was discontinued after the war because the engine was judged to be insufficiently reliable. In 1917 Glenn Martin, unsatisfied with the association with

Fig. 6: Below, a Curtiss OX-5, a liquid-cooled 90 HP (0.21 HP per cu. in.) V-8 produced starting in 1915. It powered the American Curtiss JN-4 (Jenny). At right, the nose of a JN-4, showing the radiator for engine cooling (Engine: Smithsonian Institution, National Air and Space Museum, JN-4 photo by author at the Museum of Flight, Seattle, Washington).

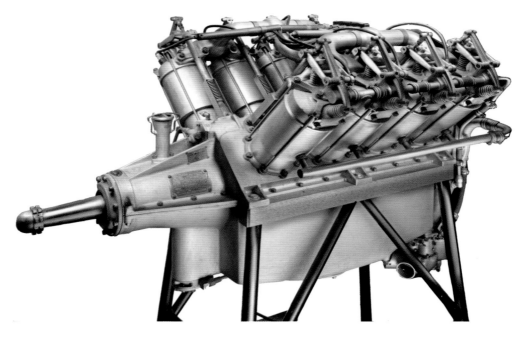

Fig. 7: A 1919 (Wright-built) Hispano-Suiza liquid-cooled V-8 with its overhead cam cover removed from the near side. (Photo by author, Museum of Flight, Seattle)

the Wright organization, had left Wright Martin to build aeroplane on his own. Thus, in 1919, Wright-Martin became Wright Aeronautical under the leadership of Frederick Rentschler and his technical chief George Mead, with the intent to enter the business of producing aircraft engines.

In addition to water-cooled engines, air-cooled rotary engines were also common on European military aircraft. Their interesting history is told below but it is brief. Air-cooled rotary engines were never mass produced in the US as their limitations were recognized.

As the US entered the war in April of 1917, it became apparent that a standard engine design for US military aeroplanes was needed and work got underway to firm up a design for easy manufacturing as soon as possible. The result was a July 1917 contract award for a 'U.S.A. Standardized Aircraft Engine' that was renamed the 'Liberty' engine. This engine was to be a family of engines with as many common standard parts as possible. The intention was to enable the construction of 4, 6, 8, and 12 cylinder engines, though only the V-12 design was built in large numbers. Design, construction, and testing took place in the remarkably short time of three months. The men responsible for the realization of this engine were Jesse G. Vincent of the Packard Motor Car Company, Elbert John Hall of the Hall-Scott Motor Car Company, and Edward A. Deeds.

As a rather narrow V-design of 45 degrees, contrasting the more conventional 60 degrees, or even the 90 degrees in the Hispano-Suiza V-8 shown in Fig. 7, the Liberty model L-12 engine was well suited for low-drag installations in aircraft, and was quickly developed to produce a substantial amount of power for the time, 400 horsepower. The design of the Liberty engine used technology developed by Daimler between 1912 and 1914 using welded-steel cylinders. Its most negative feature was that it was heavy. Nevertheless, the engines were built in large numbers by many manufacturers, primarily for aircraft, though not exclusively. In fact, the number of engines available after the war was so large that the US government insisted on their use in the aeroplanes contracted to its service. One such instance was the use of the engine in the early Boeing Model 40 mail plane whose evolution we will follow more closely.

One interesting dimension of liquid cooling is the choice of water as coolant in the early years. In the 1920s, ethylene glycol-water mixtures were developed and these allowed for coolant to be operated at higher temperatures than allowed for pure water while avoiding coolant freezing in the confines of the engines. Coolant temperatures of about 300F permitted radiators to be more compact. The coolant choice surely entered into the discussions of military decision makers when the relative merits of air- versus liquid-cooling were the topic of conversation.

Rotary Engines?

The Wright brothers and most other flyer builders used conventional liquid-cooled gasoline engines in their aeroplanes. That development will continue, but there was one significant sideroad on engine development at that time. That footnote to the history is the attempt at using *rotary*, *air-cooled* engines for aeroplane propulsion. This engine type is not related to the engines developed around 1951 by the German firm NSU Motorenwerke AG, based on the work of engineer Felix Wankel, though those engines are also described as 'rotary' and were intended for automobile and other applications including for aeroplanes by Curtiss-Wright.

The central idea behind the *rotary* engine in this discussion of early twentieth century aircraft engines is in stark contrast to the later more standard *radial*, also air-cooled, engine. In the conventional radial engine, the propeller is driven by the rotating crankshaft driven by pistons within the cylinders fixed to the airframe.

Fig. 8: A French Gnôme rotary engine. Air entry is via a central pipe intake, directly into the bottom of the cylinders with appropriate porting. Exhaust is via valve shown at right. A wire connects the stationary distributor to the spark plugs. (Photo by author at the Museum of Flight, Seattle).

Fig. 9: Cross-sectional drawing of an early Gnôme engine. Note the large ball bearings joining the portions of the engine fixed to the airframe (identified as 'anchorage plate' and drawn as a wooden structure) and the rotating piston assembly. Note attachment of wooden propeller at left. This engine has the inlet valve in the piston head. (Drawing is taken from Page (1919), see ref.)

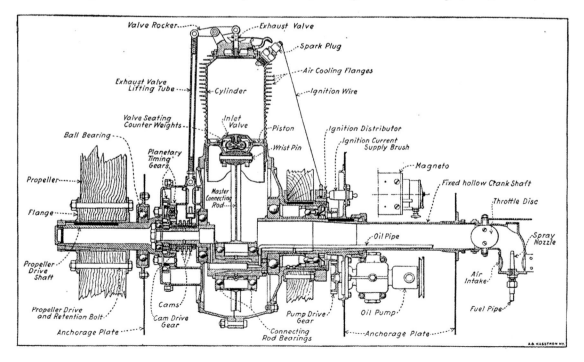

In the rotary engine, on the other hand, the crankshaft is fixed to the airframe and the cylinders rotate with the propeller. These engines were an early and effective attempt to use air-cooling in aircraft engines. Compared to water-cooled competitors, these engines were relatively light in weight.

The idea originated with an American company named Adams-Farwell between 1896 and 1905 and was carried further by two French brothers, Louis and Laurent Seguin, who introduced the concept to the aviation community about 1908. Their Gnôme engine company (Societé des Moteurs Gnôme) had been producing conventional gasoline engines for road vehicles. Another French company, Société des Moteurs Le Rhône, also went on to produce the rotary engines as their performance was appreciated for aeroplanes.

At the time of these engines' development, it was believed that, to avoid the necessity of a heavy flywheel, an advantage could be gained by having the piston-cylinder assembly act as its own flywheel. This smooth and vibration-free engine type was developed further in France, Britain, and Germany as the technology was shared before the hostilities of war broke out. Consequently there were military aircraft using rotary engines on both sides of the conflict. The chief advantage of these engines over conventional, water-cooled engines was relatively high power to weight ratio. The 7-cylinder (45–47 HP) Gnôme used during the war period weighed in at 3.3 lbs/HP, but others were very much improved. To have a smooth production of power, these four-stroke engines had an odd number of cylinders. Two complete turns of the engine resulted in a complete engine cycle.

Their design is interesting in that fuel is introduced into the engine through the central (stationary and hollow) shaft with the intake air, primarily by spraying the fuel into the airstream entering the engine. There was no carburettor. The fuel-air mixture is admitted to the cylinders of the engines by either a valve in each piston-head (see Fig. 9) or by ports near the bottom-dead-centre of the stroke. In the Gnôme engines, the valve in the piston head is opened and closed by the motion of the piston and by movable counterweights associated with the valve(s) and the pressure forces on the piston head. This design presented problems in that the valves are not readily accessible for service and the Seguins devised an ingenious valve mechanism enabling both intake and exhaust processes with a single pushrod in 1912. In this engine, the fuel is introduced via porting when the piston is near the bottom of its stroke. The engine was subsequently termed the 'Monosoupape' (translation: single valve) and the approach was copied and improved by others. In the later Le Rhône and the German Oberursel engines, the fuel-air mixture is introduced at the top of the cylinder by large tubes. See for example Fig. 10 showing such an engine, albeit without the details of the valve actuation.

Aviation is All the Rage 27

Fig. 10: A German Oberursel UR-2 rotary engine used in WWI and rated at 110 HP. Oberursel built Gnôme engines under licence since before the war. Note the tube connections to an intake manifold. Valves at the top of the cylinders control inlet and discharge exhaust. The cylinders' cooling fins were circular to accommodate the cooling airflow arriving from a variable direction: a combination of rotation speed and flight speed. (Photo by author at the Museum of the United States Air Force, Dayton, Ohio)

Fig. 11: Rotary engine installation on a Fokker E.III (Eindecker, 1915). The engine in service was an Oberursel UR.1 although this Museum of Flight model carries a Le Rhône engine. Note the open cowl at the bottom to shed exhaust under the aeroplane.

Lubrication was a challenge as oil had to be introduced into a central intake manifold with the incoming fuel/air mixture. The centrifugal forces on the oil in the rotating cylinder array led to accumulation in the cylinder heads from which it subsequently left the engine with the exhaust. The rotary engines were, therefore, heavy consumers of oil. A 100 HP two-row (fourteen cylinders) Gnôme engine might use about a gallon of oil per hour, at times as much as 2.5 gallons per hour. To put this in some perspective, the engine fuel consumption rate was of the order of twelve gallons per hour. The lost oil unfortunately bathed the pilot in the fumes. A white scarf was helpful in keeping goggles clear, and the lungs and stomach free of the oil!

Complicating the oil situation further was that the oil had to be of a type that could not be diluted by the ever-present gasoline. Such dilution would have severely reduced the oil's efficacy as a lubricant. The castor oil meeting the fuel-mixing requirements had the undesirable property of being a laxative if ingested. The pilot sitting behind the engine in an oil-laden atmosphere surely took some of the oil in. Flight endurance might not necessarily have been determined by the quantity of fuel in the tank! To be accurate however, aeroplane-designers dealt with the pilot oil exposure problem by surrounding the rotary engine in a cowl that (mostly) kept the exhaust away from the pilot by collecting it and discharging it below the aeroplane. The cowl might even have helped reduce the aeroplane drag, but that benefit was not necessarily appreciated at the time.

Lastly, it is noteworthy that the rotary engines had little or no way to modulate fuel flow; they were typically run at full power or turned off. That makes the process of coming in for a landing at least interesting. In landing, the pilot brought the aeroplane in by gliding with power off and occasionally turning fuel on to restart the engine in short bursts, termed 'blipping'. The engine could not be allowed to come to rotation rest because then it could not be restarted in flight. Blipping gave the pilot some control of the aeroplane during landing. In old movies with sound, a landing with interrupted bursts of power applied is an indication that the engine is a rotary.

The chief disadvantage of the rotary engine was dealing with the gyroscopic forces on the aeroplane inherent in the nature of the engine. There was much rotational inertia wrapped up in the mass of pistons, cylinders, and the propeller rotating at 1,000–1,200 RPM. Such inertia required significant forces to be exerted to change the direction of the rotating mass. That same inertia in a bicycle wheel keeps the bicycle upright, as does the rotor of a gyroscope always pointing in the same direction. The varying ease or difficulty of changing flight direction was a factor that pilots had to deal with. Good, well-trained pilots were, however, successful in mastering the control of the rotary-powered aeroplanes.

In spite of effective use in the First World War, these engines were never considered seriously for any later and heavier aeroplane type. Their fuel and oil consumption rates were high and the engines had no growth potential because any increase in the displacement would invariably result in a larger rotating mass making the problem with gyroscopic forces worse. Engine builders could not break the 100–200 HP barrier with them. By the end of the war, the obsolescent state of this engine type was well recognized. The 100 HP Gnôme used 9 HP just to overcome the drag losses associated with the rotating cylinders. The technology was at a dead end. The more conventional engines, at that time primarily liquid cooled, were showing more promise. Rotary engines did, however, prove that cooling the cylinders with air was practical.

The reader may have an opportunity to see rotary and radial engines in a museum or in real life and may be faced with distinguishing between them. Generally speaking, if two pushrods per cylinder are visible, it is a conventional radial engine. If only one is present, it is a rotary engine. See for example, Fig. 8. In the Oberursel rotary engine shown in Fig. 10, the intake, as well as the exhaust, valves are on top of the cylinder and the single pushrod is obscured by the cylinders. There were engines that had intake and exhaust valves on the top of the cylinder, actuated by cam driven pushrods. The French Clerget engine is a rotary with two sets of pushrods. Another distinguishing feature of Clerget engines was the use of aluminum alloy pistons.

A closing note on the fate of the French manufacturers of rotary engines. The Le Rhône and Gnôme companies were united as Gnôme-Le Rhône in 1914. After the war, when interest in the rotaries vanished, they continued as an important component of the French engine community producing Bristol Jupiter engines under license. During the Second World War, while France was occupied by Germany, Gnôme-Le Rhône produced engines for the Luftwaffe. After that war the company was nationalized and integrated into the SNECMA[10] organization of which more will be said in connection with jet engines. In short, the German engineers who were in the process of developing the BMW 003 jet engine were also added to SNECMA under the leadership of Hermann Oestrich. The author's father joined that group in France in 1946. In time, SNECMA developed a number of military jet engines in the post-war years as well as commercial turbojets (for the Concorde) and turbofans.

Air Cooling High Power Engines

Conventional liquid cooled engines were heavy and the thought developed of lightening them by cooling with air instead of water. The Lawrance Aero Engine Company, founded by Charles L. Lawrance (1882–1950) in 1917, made the first attempts at building an air-cooled engine, with (US) navy and army interest. The company started with a 3-cylinder radial L-1 producing 60 HP. The cylinders of this engine were situated at 10, 2, and 6 o'clock on the engine and appeared like a 'Y' when viewed from the front. Some installations had the 'Y' inverted. A follow-on contract with the US Army and the US Navy to build a J-1 model, producing 200 HP, was designed as a 9-cylinder radial engine with a 787 cubic inch (12.9 litres) displacement. Interest in these engines was sufficiently great that the military 'encouraged' the sale of the company to Wright Aeronautical in 1923. With this acquisition, Wright got into the air-cooled aircraft-engine business in which it did well in the decades ahead. The J-1 engine broke through the perceived belief that air-cooled engines could not be built to produce more than about 150 HP. The engine series was ultimately developed to the J-5 (also known as the R-790, Whirlwind).

---

10. Société Nationale d'Étude et Construction de Moteurs d'Aviation, or National Society for the Study and Construction of Aviation Motors.

Fig. 12: At left is the Lawrance J-1 that first ran in 1923 and evolved into the 220 HP Wright R-790 Whirlwind, at right. The display models are at the New England Air Museum and the National Museum of the United States Air Force, respectively. (Photo by author).

The postwar years also saw a number of air-cooled engine manufacturers supply engines for civil and military aircraft. Names such as Franklin, LeBlond, Szekely, Warner, Lycoming, and Continental can be listed among the many industrial concerns that manufactured aircraft engines in the period between the two world wars. Most did not participate in the development of high-power engines and many did not survive in the industry as engine builders after the Second World War. Today, Lycoming and Continental still manufacture engines for civil aviation, even though they have changed corporate ownership over the years.

The history of reciprocating engines records that C.F. Taylor had a hand in developing this engine, specifically in providing engine advice on Lindbergh's flight to Paris. Taylor was certainly one to have written the book on the technical aspects of reciprocating engines, partly through his students at the Sloan Automotive Laboratory at the Massachusetts Institute of Technology that he founded in 1929. His enduring book on the subject is included in the bibliography.

Two air-cooled engines were produced by Great Britain after the First World War. One was the Bristol Jupiter, a 9-cylinder engine producing 400 HP from 1,753 cubic inches (1.75 lbs/HP). The other was the Armstrong-Siddeley Jaguar engine, a 1,512 cubic inch two-row 14-cylinder configuration of over

350 HP (2.5 lbs/HP). These two engines were widely used in the 1920s in Great Britain.[11]

The state of the art in the early 1920s was as represented by the J-1, the Jaguar, and the Jupiter, with about .24 HP per cubic inch of displacement. That was a sixfold improvement over the Wright Flyer's engine. We will look at this parameter again later in our story when we reach the end of the era of large piston engines.

The end of the First World War left a lot of American aeroplanes and engines in private hands, sold by the government as surplus. In this environment of readily-available low-cost engines, only three US manufacturers were left able to build new ones for a profit: the Curtiss Aeroplane Company, the Wright Aeronautical Corporation, and the Packard Motor Car Company. The surplus aircraft made for difficulties for companies that focused on new aeroplane construction. In fact, Wright Aeronautical was not successful in the construction of aircraft. The company built a number of models, none of which were mass produced and none found application as significant transport aircraft. Wright followed a more profitable route with engines, but even in that domain, difficulties lay ahead.

During the war, Glenn H. Curtiss had founded the Curtiss Aeroplane and Motor Company, a prolific manufacturer of military aircraft. In 1930, it would merge with Wright Aeronautical to become Curtiss-Wright. This merger was highly ironic considering that Glenn Curtiss and the Wright brothers had waged lengthy legal warfare over the patent rights claimed by the Wrights.

In the early 1920s, the US military saw the need for new, more powerful engines and asked the available industrial concerns to develop a new air-cooled engine in the 400 HP range. The Wright J-1 was to be improved with a model R-1 (that later came to be called the R-1454) using, among other improvements, 'rotary induction', that fed a more uniform air-gasoline mixture through a centrifugal blower.

Pratt & Whitney is born

Difficulties with this 'rotary induction' effort led to a new request for proposals from the US Army. Curtiss won the competition because it was less costly than the Wright bid. The Wright engine also had development difficulties as late as 1926. The R-1454 was never mass-produced by either Wright or Curtiss. Acrimony at Wright caused Frederick B. Rentschler to quit as president and form the Pratt &

---

11. This early Jaguar and Jupiter data is from Schlaifer p.157. It conflicts with other data. The conflict is probably related to the ratings used and the model numbers produced later.

Fig. 13: The crank of a single-row radial engine. The upper case has been removed and it is concentric with the output shaft (pointing upwards). The master rod is at the 9 o'clock position with the counterweight pointing roughly towards 2 o'clock. (Photo by author at the Museum of Flight).

Whitney Aircraft Company (1925) using the reputable Pratt & Whitney Machine Tool Company as a start. Rentschler's disagreement with Wrights' directors was over their unwillingness to reinvest earnings in product development that Rentschler thought necessary for a viable future. Wrights' board of directors was primarily composed of bankers and their interests were closer to the financial dimensions of the enterprise. Wrights' chief engineer George J. Mead and A.V.D. Willgoos accompanied Rentschler and proceeded successfully to produce a lighter engine in the 400 HP size. Their main goal was to have a functional engine before the US Navy fixed on an engine called the Wright P-1 or P-2 as a standard. The development of the P-2 was led by Charles Lawrance, promoted to head the company when Rentschler left. A Pratt & Whitney concern was the US Navy's high regard for Lawrance as a designer. Pratt & Whitney succeeded nevertheless with two features: a forged, rather than cast, crankcase and the 'use of a two-piece crankshaft so that a solid master rod could be used'.[12] This allowed a higher rotational speed and with a

---

12. Schlaifer, p. 189.

greater displacement than the Simoon, the resulting Pratt & Whitney Wasp was a more powerful and technically superior engine (1926).

In addition to military sales and deliveries, forty of the Wasp engines were delivered in 1927 for installation on the Boeing Model 40A Mailplane. When the Boeing Mailplane was designed in 1925, the original Model 40 was built with water-cooled Liberty engines with which it was underpowered. At the time, there had been government 'encouragement' to make use of these surplus engines. The decision to refit the Model 40 with Pratt & Whitney engines provides a hint that Boeing would be further involved in vertical integration of industrial functions where the manufacturer of aeroplanes and the necessary components also operated them. Indeed, such an event would occur in 1929.

Contract Air Mail

After the First World War, the US Post Office was the one commercial customer for whom transport services were provided by people with aeroplanes. Initially the conveyance of mail was an informal arrangement, but with passage by the US Congress of the Contract Air Mail Act of 1925 (also known as the Kelly Act), the relationship between flyers and the Post Office became more structured. The Kelly Act allowed the Postmaster General to enter into contracts with private air carriers, based on bids to provide such services. Routes were awarded to qualified flyers who, for a fee per pound of delivered mail, connected a number of important American cities with their services. The goal was rapid delivery of mail. In the following four years, 34 C.A.M. routes were established. Among the individuals to take advantage of the opportunity was William Boeing who, in February 1927, created the Boeing Air Transport Company for the purpose (among others in his future) of securing C.A.M. route no. 18 between Chicago and San Francisco. His successful bid, and that of the many others who went on to establish notable reputations as aviators, set the first significant commercial aviation adventure in motion.

The marriage of aviation and mail in the 1920s and '30s is an interesting chapter in the development of commercial aviation. It was at the centre of the early attempts to establish safe air transport across the country. It was also a turbulent time when viewed from the perspective of service to the public. The evolution of the mail transport and the nascent passenger business in the US was determined to a large degree by an act passed by Congress that gave the Post Master General authority to award the mail hauling contracts. The goal of the Post Office was to keep the number of the airlines it dealt with small and their efficacy reasonably good. The Air Mail Act of 1930 led to a meeting of major airline operators who divided the country into regions where they would serve the Post Office, in effect keeping others from bidding for the government contracts that were still a major element of profitability. This meeting was later described as the 'Spoils Conference'. The meeting focused power in four major airlines that were to play critical roles in the growing airline business in years to come. These carriers were United, American, and Eastern Airlines as well as TWA (Transcontinental and Western Airlines) that had recently been created by a merger of two smaller airlines. These airlines were keenly aware that the stock market crash of 1929 would require them to take drastic steps to insure survival.

The possibility of shadiness in the route awards to the flyers ultimately brought the matter to the public's attention where it became the 'airmail scandal'. The telling of the fascinating and entangled story of the Post Office, the US Congress, US government administrations, and the aeroplane operators is left to others. In short, the Congress wrote new legislation with consequences for some members of the aviation industry that were dire indeed.

The First Paying Passenger

The Boeing Model 40A, with its Wasp engine, had the capacity for carrying two people, perhaps mechanics or deadheading pilots. When this functional need was not pressing, paying passengers could be and were carried. With the good publicity

Fig. 14: A Boeing Air Transport Boeing Model 40A with accommodations for two passengers flying the Contract Air Mail route No. 18 between San Francisco and Chicago. Photo taken flying near Ruby Mountain in Nevada. (Boeing Images BI210144).

of travelling by air, came the motivation to do better. The later Model 40B-4 (a close relative of the Model 40C) was redesigned to carry four passengers. The principal function of the Model 40 was, however, to carry up to 1,200 pounds of mail.

In contrast to the location of the mail and passengers who were seated just behind the engine in the fuselage with closed doors, the pilot sat in an open cockpit at the rear. In those days, the open cockpit served to keep the pilot well informed of the state of the aeroplane, its engine, and the environment, the latter for atmospheric conditions, icing in particular. He also needed good visibility of the ground so that he could navigate over the countryside. Variations of this aeroplane, which could be called the first airliner, were produced between 1927 and 1932. Boeing produced about eighty of them.

The success of the Pratt & Whitney Wasp moved Wright to produce the R-1750 Cyclone, with 500 HP. The stage was thus set for further developments in air-cooled engines in the US and a more limited use of liquid-cooled engines in civilian and military aircraft. The industry now had the capacity to build lightweight and reliable air-cooled engines. For example, the Wasp B (1927) weighed in at 1.49 lbs/HP. Compared to the Liberty weight at 1.88 lbs/HP, significant progress had been made in just a few years.

Big Engines in Germany

The Allies started to relax the Versailles Treaty obligations on Germany in the late 1920s. Germany was permitted to produce small air-cooled engines for sport purposes and larger ones under licence for commercial purposes. In 1928, BMW was licensed to produce the Pratt & Whitney Hornet A engine (9 cylinders, radial) delivering about 510 HP. BMW was further licensed in 1933 to produce the Hornet B, the engine that found its way into the Junkers Ju52 commercial transport. The company took this exposure to American technology to develop air-cooled engines on its own terms by building the BMW 132 in two versions: a carburetted version for civilian applications and a fuel-injected version for the military. After 1933, the new Nazi government ignored the Versailles Treaty altogether and funded development of its military, including, of course, the necessary engines.

The BMW 132B engine was ultimately pushed to produce 1200 HP and that performance limitation led to the design of a 14-cylinder (two rows of seven) engine known as the BMW 139. It first ran in 1937 and was, by that time, developed without much interaction with Pratt & Whitney as the political situation between Germany and the rest of the western world was deteriorating.

To a significant extent, interest in aircraft engines by German builders was focused primarily on liquid-cooled engines, though not exclusively. In the 1920s and 1930s, the engine-building firms Junkers Motorenbau, Daimler Benz, and BMW, among others, developed innovative high-performance engines, initially for commercial purposes and later for use during the war. It is safe to say that the last of these engines played little to no role in the evolutionary history of transports and airliners that was to follow the end of the war, which halted any possibility for further engine technology developments in Germany.

Towards Good Aircraft Engines

A historical perspective of the engine industry yields insight into its complexity. Many avenues were examined and many were rejected. The building of engines has always been a matter of choosing from a number of ways of achieving the goal of a 'good' rather than a 'perfect' engine. Some options, such as the rotary engines, were dead ends. Other engine types might be satisfactory for some purposes but not necessarily for aviation. The reality that some engine types never found widespread application in aviation is a measure of that judgment. For example, two-cycle internal combustion engines and Diesel engines were seriously examined but never adopted for use in large commercial aircraft.

When there are two ways of building a good engine, then preferences by engineers and customers come into play. For the decades of the dominance of the internal combustion engine, the choice between water cooling and air cooling separated the industry into 'camps'. When the gas turbine became a reality, the question of air compression via radial flow or axial flow compression was open. That question has been settled with improved understanding of the physics involved.

The Parallel Evolution of Wings

The ongoing need for engines of greater power output during the first two or three decades of aviation was driven, in large part, by the drag experienced by the aeroplanes as they were made to fly faster. The wings of the day were heavy contributors to that drag. From the Wright brothers onward, to the mid-1920s, biplanes were the common design. The requirement was for a lightweight and stiff beam structure to generate the lift force of the wing, without unwanted deflections, especially twists, that could affect the airflow over the wings.

The low speeds of the day, imposed by power limitations, the desirability of low landing speeds, and the not yet wide use of wing flaps, required the area of the wing to be substantial to generate sufficient lift. The span was materially reduced in the biplane design, hence its practicality.

From a structural viewpoint, one may think of the biplane wing as a box beam with the two webs removed and replaced by struts and wires. The struts were there to hold the upper and lower surfaces apart, while a plethora of diagonal wires provided the necessary stiffness. The array of wires was so extensive that their drag was an important part of the total drag of the aeroplane. Further, there was a drag penalty associated with interference between the wings which led innovators to place one wing (usually the upper) forward of the other. This practice is called *staggering* the wings. The evolutionary end point of multi-plane wings was the use of an upper wing significantly longer in span than the lower. Such an aircraft plane is called a *sesquiplane!* The Boeing Model 80[13] could be classified as such a design even if its lower wing was somewhat larger than half its upper wing.

The materials available to the early builders were wood, fabric, and steel wires. Steel fasteners of various kinds were used, but wholesale use of steel for wing structure was prohibitive from a weight perspective. The design of wings was made much easier with the wide and cost-effective availability of aluminum as a structural material in the mid-1920s. It was light and strong. Aluminum had been available since the late nineteenth century, but it took time to develop it into useful alloys and for the industrial production infrastructure to be generated. In the aeronautical world, aluminum allowed building the wing with sufficient thickness so that the spar and ribs could be enclosed within the skin and, significantly, no external structures were required. With its introduction into the construction technology, wooden spars, ribs, and fabric coverings disappeared from the new aeroplanes. The monoplane was born and it prospered.

The switch from biplane to monoplane changed the nature of the drag experienced by the aeroplane. The parasitic drag associated with the structural members of the biplane wing disappeared and was replaced by the ever-present drag due to lift. Skin-friction (also called parasitic) drag of the aeroplane body and wing surfaces was still important, but engine power could now be employed to increase speed. And indeed, it was. The drag due to lift is called *induced* drag, because it is induced by the wings' forcing a downward airflow and creating the associated wing-tip vortex. The trailing vortex system in effect causes the aeroplane to fly continuously *uphill*, which is the ultimate cause of induced drag. This important source of drag was keenly investigated by aerodynamicists during the early 1930s. Their work led to identification of wing planform, that is to say, wing shape, as a part of the design to minimize drag. Such aerodynamic work resulted in the use of an elliptical shape

---

13. The Boeing Model 80 was a 12-passenger airliner in service with United Aircraft and Transport Company from 1928–34. A later Model 80A with Pratt & Whitney Hornet engines carried 18 passengers. A total of 16 aeroplanes were built.

of a wing to achieve a uniform downwash. The English Supermarine Spitfire is a good example of such a wing design and there were many others. It was soon apparent, however, that a tapered wing was almost as good and less costly to build. Hence it dominates that aspect of wing design.

Thus, not only did engines determine what an aeroplane could do, but its wings also evolved for a better combination of thrust and drag to allow for higher speed.

New Engine Directions

Engines for aircraft were made ever more powerful and allowed the aircraft to grow in capability. The decade of the twenties was also when the steam turbine became an increasingly important part of the infrastructure to produce commercial electric power. Publications such as the engineering text of Aurel Stodola (*Steam and Gas Turbines*, 1923) allowed engineers to build ever-larger steam turbines and to consider the use of ordinary gases like air and combustion gas to be processed by rotating machinery.

Another important contributor to the ideas central to turbomachinery was A.A. Griffith (1893–1963), an English engineer who suggested early on the possibility of the axial flow compressor in a 1926 paper entitled 'An Aerodynamic Theory of Turbine Design'.

In the 1920s the Swiss firm Brown Boveri sold engines that resembled a modern gas turbine (or jet) engine. These were special purpose engines for emergency power and for applications where a lot of hot air was needed as, for example, in the boiler of a steam power plant. These engines were poor performers in terms of overall efficiency, but they worked to convert a few percent of the thermal energy in the fuel to mechanical power by the same means as would later be used in the jet engine. The mediocre performance level was due to compressor and turbine efficiencies, as well as the turbine inlet temperature that were all low, when viewed from today's perspective. Turbine inlet temperature was strictly set by the limits imposed by materials used in the turbine as no blade-cooling technology was yet available. These parameters would markedly improve in the years to come, leading to the notion of using the basic ideas for an aeroplane-propulsion engine.

Supercharging and Turbocharging

Piston-aircraft engines were always demanding greater amounts of power for desired performance improvements. As the technology of both expanding and compressing gas by dynamic, or better said, aerodynamic, means was mastered, it was logical to explore the use of a supercharger. Such a compressor increases

Fig. 15: Left: A cutaway GE turbocharger. Engine exhaust enters the top opening and exits via the axial flow turbine to the left. The compressor is radial with an air collection scroll around the rotor outlet. Air enters at the right flowing towards the impeller to the left. The exit of the high-pressure air is not visible. (Courtesy General Electric). Right: installation of a turbocharger on a B-17. Note waste gate valve in the exhaust duct to control the compressor output pressure. (Photo by author at the Museum of Flight).

the density of the intake air, allows more fuel to be burned, and thus increases the power output from an engine. The charging works particularly well at higher altitude. However, it does use some of the mechanical shaft power, and a somewhat better approach is to use the otherwise-wasted power in the exhaust stream of the piston engine by means of a turbine to drive the charger. A device using this approach is called a turbocharger. The developments of these chargers strengthened the understanding of the basic elements necessary for the invention of the gas turbine in the next decade.

The idea of turbocharging an engine for operation at high altitude was a Swiss idea by Alfred Büschi (1901) and adapted to World War I aircraft by the Frenchman Auguste C.F. Rateau with measurable success. As the US entered that war, details of this technology were shared with France's new American ally via the National Advisory Committee on Aeronautics, NACA, founded in 1915. The General Electric Company (GE) took a keen interest in the technology because of their related experience with steam turbomachinery. The company, under the capable direction of Sanford A. Moss,[14] tested the super/turbocharging concept in the

---

14. Moss received US Patent 161214A (assigned to General Electric) in 1925 for the idea of a turbocharger, unaware of a similar French patent publication in 1921 by Rateau.

rarefied atmosphere of the Rocky Mountains and on aircraft engines. In the 1920s GE turbochargers found their way into many engines powering aircraft.

An interesting aspect of supercharging and turbocharging is that the air compressed for delivery to the cylinders of the internal combustion engine rises in temperature as a result of being compressed. That, unfortunately, reduces the density of the air and reduces the amount of fuel that can be added there. A common means of dealing with this is to cool the air by means of a heat exchanger after turbo-compression and thus restore some of the air density. This is known as 'intercooling' and would be used in many later high-performance engines, including those used in the Boeing B-17.

An eventful Year

The year 1929 saw a number of significant events. In February, William Boeing and Frederick Rentschler (founder of Pratt & Whitney) formed a vertically-integrated company that united business interests in all aspects of aviation by combining engine and airframe manufacturing and the airline businesses with United Airlines as the centrepiece. The goal was to serve all aspects of the aviation market. The civil aviation portion was to include cargo, passenger, and airmail services. On the military side, business opportunities for responding to military aviation needs were also to be followed.

The new holding company, the United Aircraft and Transport Corporation (UATC), with headquarters in Hartford, Connecticut, controlled the stock of the Boeing Aeroplane Company, the Chance Vought Corporation, the Hamilton Aero Manufacturing Company (a propeller manufacturer), and the Pratt & Whitney Aircraft Company. Sikorsky Aviation Corporation, the Stearman Aircraft Company of Wichita, Kansas, and the Standard Steel Propeller Company were added to the corporate portfolio in short succession, as were several more airlines.

Two other important events in 1929 had consequences for this aviation story. One was the American engine industry reduction of its number of producers of large air-cooled aircraft engines from three to two. There might not have been enough business to support all three. In July 1929, Wright and Curtiss joined forces to become Curtiss-Wright to more effectively compete with Pratt & Whitney.

Finally, October was to bring financial mayhem to many in the form of a stock market crash, including the airline industry. In the years of the Great Depression that followed, the industry was slowed, but survived.

Chapter 3

The 1930s: Real Transport Aeroplanes

Air Travel Infancy

The initial experience of carrying passengers in the Boeing Model 40 led the company to the building of larger aeroplanes to carry more passengers. The airliners of that time were still, as one might expect, small. A Model 80 was designed in 1928 shortly after the creation of the Boeing Air Transport Company in February 1927. A year later, the Model 80 (12 passengers) was quickly replaced by a larger Model 80A (18 passengers) with a take-off gross weight of 17,500 lbs. The Model 80A was powered by three Pratt & Whitney (R-1690) 575 HP Hornet engines and had a wing with wooden spars and fabric cover. The sixteen examples of the Model 80, most of which were 80As, were built specifically for Boeing Air Transport Company and retired from company service in 1934. The six-year life for this Boeing model was indeed short, in part because technology was rapidly being developed that allowed the design of better airliners.

Other representative airliners of that era were the Ford Trimotor, the Junkers G24, and the Fokker F.VII.[1] These monoplane aircraft were lighter and smaller than the Model 80A. They weighed about 10,000 lbs at take-off and carried 8–12 passengers. Cruise speed was a little more than 100 mph and the range was about 500 miles. These numbers pretty much describe how these airliners were used: short hops, primarily overland. The innovation introduced with these three aircraft was that they were closer to all-metal construction and included few, if any, fabric surfaces that were the norm in that day. To illustrate the impact of an all-metal aeroplane, consider that a fabric-covered aeroplane (a de Havilland Dragon, for example, produced in the next decade) with the same capability weighed about half as much as the all-metal design. That difference emphasizes the need for greater engine power for all-metal aeroplanes.

---

1. This three-engine aeroplane was of steel tubing and plywood skin construction and was produced between 1925 and 1932. It was a good competitor to the Ford aeroplane. Its builder, Anton Fokker (1890–1939), built fighter aircraft for Germany during the First World War. He moved his business to Holland because of limitations imposed by the Versailles Treaty and created the Atlantic Aircraft Corporation in the US.

Fig. 16: Ford Trimotor of which about 200 were built between 1926 and 1933. Note the absence of any cowling over the air-cooled engines. (Peter M. Bowers Collection/Museum of Flight).

The Ford Trimotor was the result of an investment in the Stout Metal Aeroplane Company whose owner William Bushnell Stout designed a number of aircraft including the Trimotor. Ford eventually bought the company and produced the Ford 4-AT (and later models) Trimotor between 1926 and 1933. About two hundred of the aeroplanes were built until competition from more capable aeroplanes like the Boeing 247 and Douglas DC-2 rendered the design obsolete. The Trimotor design, seating eleven passengers, was based to a significant extent on the work of Hugo Junkers with a visual similarity in that both builders used corrugated sheet metal skins. The advantage of corrugation was the stiffness of the sheeting did not require as extensive a frame and longeron structure as a flat sheet would require. Naturally, because the external air flow did not necessarily line up with the corrugations, there were undoubtedly drag penalties incurred by this construction technique.

None of the aircraft of that day had pressurized cabins. That fact limited them to flight at relatively low altitude at the mercy of the weather that was always a major aspect of flight operations. The little bags in the seat pocket were quite likely to be used for returning the passenger's last meal to freedom when the flight was rough. Similar little bags may be found in the seat pocket of a modern jetliner but, mercifully, they are hardly ever used.

Big Radial Aircraft Engines

The piston engines of the day were of two kinds: air-cooled and liquid-cooled. The more common air-cooled variety generally involves a design with a number of cylinders arranged in a radial pattern (typically 7 or 9) around the propeller shaft so that all cylinders have good access to cooling air from the oncoming airstream.

Alternatively, the liquid-cooled (usually water with antifreeze) variety can be designed as an in-line or 'V' design, for good streamlining of the aeroplane or engine nacelle. But it requires the use of a radiator and necessarily involves drag associated with it. The liquid-cooled 'V' configuration was not unlike an automotive V-8, often with more cylinders. For the power demanded by the bombers and transports of the day, liquid-cooled engines were uncommon in the 1930s. They were employed, however, in a number of later military fighter aircraft, *e.g.* the P-38, P-39 and the P-51.[2] In the US, transport and bomber aircraft used air-cooled piston engines almost exclusively. The reason is that less complexity is involved: no coolant, no antifreeze, no pumps to fail, and no coolant leaks!

It is no exaggeration to say that this was the heyday of mechanical engineering as a discipline in which understanding of mechanics and materials was necessary for the building of a good product. Consider the engine as it executes the Otto cycle elements of intake, compression, combustion, expansion, and exhaust. The configuration that allows these processes to be handled successfully is highly complex and required technical innovation on many fronts.

Radial engines have an odd number of cylinders. This notion was arrived at by the four-cycle nature of the piston motion. In these engines, the arbitrarily identified odd cylinders (1, 3, etc.) might sequentially execute the intake stroke while the even ones execute power strokes. Finishing the rotation will have the odd set compressing the air while the even ones force the combustion gas out of the cylinders. Another revolution of the central crank will result in a complete execution of the cycle by all cylinders. The dividing of cylinders into 'odds' and 'evens' by devising an ignition sequence makes the power or torque output relatively smooth.

One might ask: with an odd number of cylinders won't there be more of the odd than the even cylinders? The answer is that the odds and evens switch roles at the end of each revolution, so that after two revolutions the number of each will be the same.

---

2. The US military designation 'P' is for pursuit, changed after the Second World War to 'F' for fighter.

This ingenious configuration allowed the construction of lightweight radial engines.

The field of mechanical engineering relied heavily on the understanding of gears and cams to drive the necessary intake and exhaust valves. No mechanical engineering student in the first half of the twentieth century escaped a course in this arena. Materials and lubrication also had to be fully understood so that frictional contact between components did not lead to wear and failure.

These radial engines had to deal with the management of heat, as did all internal combustion engines. Fortunately, the exposure of wall and valves to extreme temperature gases was short in the internal combustion engine, just a few thousandths of a second, and a fresh cool charge of air followed closely behind to take away some of the heat from the affected surfaces. That characteristic and cooling fins (or liquid coolant) limited average metal temperatures to modest levels, normally around 400 degrees F or 200 C. The cylinder-head temperature was a closely watched parameter by the flight engineer or pilot. It was no small feat of the engineers of the time to design, build, and operate these engines.

Labels for Engines

Starting in 1924, American reciprocating engines included the number of cubic inches in the model name. Thus, the common way of describing reciprocating *radial* engines as a model number was by the letter 'R' followed by their displacement in cubic inches. For example, we have R-1820. Similarly, the letter 'V' is used to describe engines like the Allison V-12, as in V-1710.

The use of letters to describe engine types was carried forward into the jet age with 'J' to denote turbojet engines purchased by the military (*e.g.*, J47), 'T' turboprop engines (*e.g.*, T53), and 'TF' for turbofans (*e.g.*, TF39). In the case of these last three engines, the number after the J, TF, or T is a sequence number assigned by the military that funded the development and is not significant in any other way. The engine description 'name' for commercial jet engines is invented by, and specific to, the manufacturer. Unlike the reciprocating engines for which engine size is indicated in the number, no such parallel was established for characterizing the size of the members of the jet family. There was an early use of thrust level to characterize an engine, but that turned out to be silly as thrust level improved as engines were developed. The GE I-16 and I-40 to be described later in our story were stuck with labels of this kind.

Fig. 17a: A museum display of a Wright R-1820 Cyclone. This 9-cylinder engine type powered the B-17 and the DC-3, among others. (Photo by author, New England Air Museum).

Fig. 17b: Cross-section view of a Wright R-1820 Cyclone showing the essential features of this (single row) engine type: carburettor in the upper right, impeller for the supercharger, crank with piston on master rod at top-dead-centre, and direct (no gearing) output shaft at left. Not shown in the sketch are details of the cylinder head: valves, spark plugs, and manifolds. The valve pushrod image is terminated just above the cam. At the bottom is the oil sump.

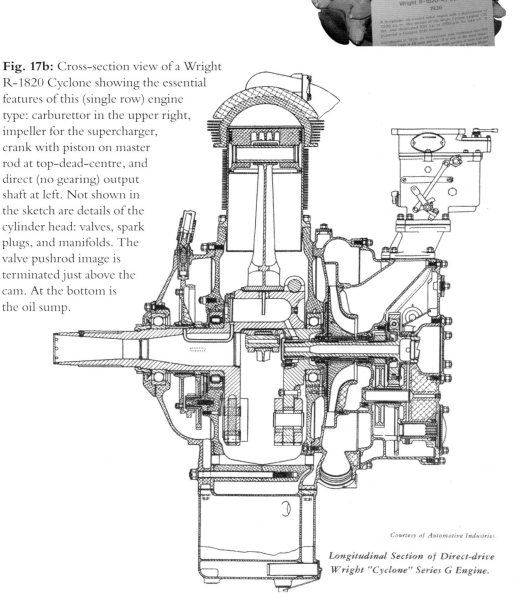

Courtesy of Automotive Industries.

Longitudinal Section of Direct-drive Wright "Cyclone" Series G Engine.

The 1930s: Real Transport Aeroplanes   47

Fig. 18: An Avro Lancaster bomber. Note the slim liquid cooled engine nacelles (Rolls-Royce Merlin V-12s) and the radiator inlets below the propeller spinner (Photo by Raymond Murray at the Canadian War Museum, Ottawa, Canada).

Liquid-cooled engines were used primarily in fighter-type aircraft. Widely used fighter aircraft using Allison V-12 engines were the Lockheed P-38 (Lightning)[3] and the Bell P-39 (Airacobra). Allison, a division of the General Motors Corporation, was an important engine builder in the 1930s and '40s. Probably the best-known US military aeroplane with a liquid cooled engine is the North American P-51 powered by a Packard V-1650 engine (1,400 HP). The engine was a supercharged V-12 design produced under licence from Rolls-Royce where it was called the Merlin. Liquid-cooled engines were used in bombers like the British Avro Lancaster (1942) and the German Ju88 (1939). The earlier British heavy bomber, the Vickers Wellington, with its fabric-covered airframe, used radial engines. No liquid-cooled engines found their way into mass produced airliners, so mention of this engine type can rest at this point.

The story of engine power during the 1920s and '30s can be described with a couple of examples. The Ford Trimotor (Ford-4AT-E) was built between 1926

---

3. The P-38 was a twin-engine aeroplane designed to be manoeuvrable. To that end, its engines rotated in opposite directions to minimize gyroscopic effects on flight.

Fig. 19: Left: A Pratt & Whitney engine dating to the 1930s: a 9-cylinder R-985 Wasp Junior. (photo by author at the Museum of Flight). Right: a Wright R-2600 Twin Cyclone. A sectioned view of this engine is shown in Fig. 23. (Photo by author at the Evergreen Aviation and Space Museum, McMinnville, Oregon).

and 1933.[4] It was powered by three Wright J-6-9 (R-975) Whirlwind engines of 300 HP and carried eleven passengers at 100 mph. The Douglas DC-3 was designed ten years later and built as an airliner between 1936 and 1942. It carried 21 to 32 passengers and was powered by two Wright or Pratt & Whitney engines of about 1,100–1,200 HP that allowed cruising at 200 mph. The Wright Cyclone was a nine-cylinder engine while the Pratt & Whitney Wasp was a 'twin' with two sets of seven cylinders. Air transportation was improving by significant steps. Looking at the numbers, one may conclude that doubling the power doubles the speed and capacity.

During the war, Pratt & Whitney would develop the Twin Wasp further to a rating of 1350 HP. That engine would be used in the next generation of Douglas transports, the DC-4.

In a 'twin' or 'duplex' engine, the second set of cylinders was orientated relative to the first so that it also had access to cooling air. Fig. 19 illustrates this configuration. With seven, rather than nine, the space between cylinders is wider, but seven and nine cylinder arrangements were used. Thus, two-row, air-cooled reciprocating engines had either fourteen or eighteen cylinders.

Leaping ahead in time to the 1940s, we note later developments brought about powerful twins. The Wright Twin Cyclone (R-2600) was a fourteen-cylinder

---

4. Dates quoted are year of entry into service to end of production.

engine while the Wright Duplex Cyclone (R-3350) and the Pratt & Whitney Double Wasp (R-2800) were built with two rows of nine cylinders. These single-row and double-row engines powered many military aeroplanes during the Second World War. For example, the Boeing B-17 had single-row Wright Cyclones while the Pratt & Whitney Twin Wasp found application in the Consolidated B-24 (Liberator). The B-17 and the B-24 had similar cruise speed capability of about 200 mph. The later B-29 was powered by significantly higher power Wright Duplex Cyclones allowing cruise speeds closer to 300 mph.

A few years later, Pratt & Whitney even developed an engine with twenty-eight (four rows of seven) cylinders, the R-4360. That engine will again play a role in our story of engine evolution.

The First Airliners: Industrial Warm-up

The 1930s saw the introduction of larger airliners. The airliners before that decade were primitive by today's standards. They were slow and small. Also there were not many of them. The power of new engines allowed the welcome replacement of single- and three-engine aeroplanes with two engines on the wings, a much more attractive design, partly because it would result in a quieter cabin.

In Europe, a number of airliners were built, chiefly in Great Britain and France. They operated not only in the countries where they were built but also in the larger world. In Germany, civil aviation was reborn after the war with the introduction of an all-metal airliner, the Junkers Ju52 (1932). This three-engine aeroplane was a very successful design produced in large numbers, although the thousands noted in the literature include units produced for the military in the Second World War. It used BMW Hornet B engines produced under license from Pratt & Whitney. The German airline Lufthansa operated these aeroplanes before and after the war and retains a flying example for publicity and nostalgia.

In the United States, a number of manufacturers of transport aeroplanes rose to the business opportunity presented by a growing airline industry, including the main subjects of our tale, Boeing, Douglas and Lockheed.

Boeing ventured into the business with a single-engine aeroplane called the Monomail. Two prototypes were built called the Model 200 and the Model 221 starting in 1930. It was sized for six passengers and powered by a single Pratt & Whitney Hornet B with 575 HP. The first use of retractable landing gear was incorporated in the design. In 1933 the aeroplane was retired from service and Boeing focused its efforts on a twin-engine Model 247 that also included this feature. The 247's design had a sleek all-metal, low wing configuration. A design goal was to substantially increase speed with aerodynamic 'cleanliness'

Fig. 20: A Junkers Ju52. Note that the skin is made of corrugated aluminium, a common construction technique when stiffness of the exterior skin is desired. This airframe was not pressurized. The photo shows the short cowl on the centre engine. (Courtesy Lufthansa Archiv).

and powerful engines to allow operation above the 100 mph capability common during the 1920s. Boeing settled for reliable engines (Pratt & Whitney Wasp with 575 HP) rather than take a risk with the more powerful Wright Cyclone, R-1820 (Fig. 17) with around 700 HP that might not be ready in time. The Wasp engines were powerful enough to drive a higher speed (170–180 mph) but only with a smaller take-off weight. Thus the 247 had to be modest in size.

Douglas entered this competition for the airliner business with a DC-1, followed in rapid succession by the DC-2. The DC-1 design and the commercial airliner construction programme on the part of Douglas was initiated in part by Jack Frye, president of TWA (more accurately T&WA). His tenure at the company as chief executive covered the years 1934–47. Interestingly, it is during the upheaval of the Air Mail Scandal that Frye made TWA into a major American airline. In 1939 he would secure financial support with Howard Hughes's investment in the company and, unfortunately, suffered departure from the company as president, largely as a result of disagreements with Hughes.

The one example of the DC-1 and the production DC-2s were configured with versions of the Wright Cyclones. The 247 and DC-1 were first flown in 1933. The 66 inch wide DC-2 started service in the following year.

American Airlines had sought to buy 247s from Boeing. The Boeing commitment to the large United Airlines order and the time lost waiting for delivery was unacceptable to American. They turned instead to Douglas with the request for a larger body diameter to accommodate 14 to 16 sleeping berths. Douglas was not necessarily enthusiastic about investing in a new aeroplane but agreed to it. Thus, the long-lived DC-3 was born. Its fuselage was 92 inches wide and had a capacity of 21 to 32 seats in a 1+2 configuration. The DC-2 and DC-3 cabins were sufficiently large to allow for wing spars to be located under the floor, resulting in a spacious cabin and unobstructed aisle. It is not hard to imagine that some airline operators would later configure the seating of the DC-3 to be 2+2 to increase the passenger capacity of the aeroplane.

Lockheed was also a competitor in this field at this time with its Model 10 Electra. This aeroplane type gained considerable fame for carrying Amelia Earhart on her ill-fated voyage around the world. In size and capability, the Model 10 was similar to the Boeing 247.

Cowlings

A pressing issue of the day was the cooling of the cylinders of the radial engines without incurring excessive drag. Cooling was carried out by heat conduction from the cylinder walls and head to metal fins exposed to cooling air. The total fin surface area was typically of the order of 100 square inches for each square inch of piston area. The earliest generation of aeroplanes had the cylinders completely in the free stream of outside air. The Ford Trimotor and the Boeing Models 40 and 80 were configured this way. That was a very 'draggy' configuration and not consistent with the desire for speed. Research by engine manufacturers and NACA led to the identification of a good cowling configuration and baffling between cylinders to optimize cooling. The baffles were necessary to direct cooling air to the rearmost parts of the cylinders' cooling fins. On the cowling issue, the argument was between use of a short cowl just outside the cylinder heads, or a longer cowl whose air outflow could be controlled. The latter proved to be the best and was subsequently adopted for almost universal use. In practice, the flight engineer or pilot would monitor the cylinder head temperatures and close or open the cowling flaps as appropriate. The early Boeing 247 had a short cowl.

Propellers

The propellers used through the period of the First World War were wooden. They were typically laminated with metal leading edges. In the early 1920s, aluminum propellers became popular. Up to this time, the geometry of the blades was fixed; a propeller of this type is said to be *fixed pitch*. As time went on in that decade, propellers with pitch adjustable on the ground became available to fulfill the needs of a particular flight. By the end of that decade, hollow steel propellers became the norm. Steel was hard and strong.

If high aeroplane speed was to be realized, propeller pitch had to be adjustable in flight. Fixed pitch places a serious limitation on a high-speed aeroplane because good thrust in cruise would involve a blade pitch that is too steep at low speed, inviting propeller-blade stall. Under these circumstances, the propeller is inefficient at take-off and the aeroplane consequently requires a long take-off roll. This situation could be readily solved if propeller pitch was adjustable in flight. In 1929, two patents[5] were filed to allow this pitch variation, granted to W.R. Turnbull in 1931 and F. Caldwell in 1937. Frank Caldwell at Hamilton Standard (and earlier when the company was the Standard Steel Propeller Company) contributed much to propeller technology. His list of patents is longer than those cited here and his work was recognized with the award of the Collier Trophy in 1933.

The Collier Trophy has been awarded annually since 1911, sometimes by US presidents, to individuals who have made significant contributions to aviation. The award is administered by the US National Aeronautic Association and is named for Robert J. Collier, an aviation enthusiast who owned a Wright Model B biplane and who published *Collier's Weekly* magazine. This award was also awarded to others in our story.

Constant speed automatic pitch control for propellers became available in 1935. The technology was quickly adopted for the faster airliners where a high efficiency propeller was needed for a wide range of flight conditions. Full feathering capability[6] of propellers was not incorporated into propellers until 1938. Finally, in 1945 propellers were built with reverse thrust capability.

---

5. US RE20283 E, 'Propeller', by F. Caldwell, published 9 Mar 1937 (filed 25 May 1929) and US 1828303 A, 'Variable pitch propeller', by W. R. Turnbull, published 20 Oct 1931 (filed 14 Aug 1929).
6. Adjustment of the blade orientation so that an engine failure in flight could at least minimize the drag associated with the 'dead' engine.

Truly modern Airliners: Round One

The DC-2 adopted the variable pitch propeller in 1933 and the 247 did it a year later. The failure to recognize the technology jump in a timely manner cost Boeing sales of the Model 247. The situation was compounded by a marketing error committed by Boeing in that the company agreed to sell the first sixty 247s to United Airlines (its corporate member airline) leaving TWA (then known as Transcontinental and Western) and others to go with the Douglas DC-2. A further blow to Boeing was that the DC-2 was of an advanced structural design whereby the Douglas engineer John (Jack) Northrop (who had by then left Lockheed) included a stressed skin in the wing structure. This idea originated in Germany during the First World War at Junkers and Rohrbach Metall-Flugzeugbau and was adopted in the US in the DC-2. The 247 was rather old-fashioned in that it carried all wing loads through the spars with the wing skin playing a passive structural role.

Fig. 21: Boeing Model 247D interior and exterior. Note the wing structural member (spar) on the floor over which one would have to step. The in-flight picture shows the aeroplane in Boeing Air Transport Company livery. (Boeing Images BI228125 and BI225469).

In the competition with the DC-2, the 247 did not fare well because it was committed to the United Airlines order and competitors introduced new aeroplanes. The 247 and the similarly sized Lockheed Model 10 Electra, introduced in 1935, were smaller than the Douglas DC-2 and even less competitive when the DC-3 began service in 1936.

American transport aeroplanes were produced in sizable numbers for American as well as overseas customers. Ultimately, the production numbers were 75 Model 247s, almost 150 Electras, and about 200 DC-2s for commercial service. By far the most successful, judged by the production numbers, was the Douglas DC-3, of which more than 450 were built for airlines before the Second World War. The DC-3 became a transport workhorse later in the war as the C-47. In terms of performance, cruise speed for the fleet of all airliners was rising to an average of 175 mph, topping 200 mph for the later DC-3. Travel times were becoming shorter indeed. Range also increased to 1,500 miles for the DC-3, which was still not enough to cross the Atlantic non-stop.

The year 1934 was not a good one for Boeing. It was the time of the 'Air Mail Scandal' when Congress stepped in to 'correct' the airmail route award process brought about by the Air Mail Act of 1930. The Roosevelt administration eventually cancelled the lucrative airmail contracts with private carriers. All this was happening just as United Airlines was taking delivery of its new 247s. United Air Transport Corporation (UATC) had planned deployment of the 247 in part for its airmail service. In the chaos that followed, mail transport was taken over by the military, a task they were not fully equipped to carry out efficiently. The service was subsequently returned to private carriers with the entire air industry under close scrutiny. When it was over, Congress drafted a law (Black-McKellar bill, 1934) that forced the breakup of the vertically integrated UATC. Portions of the company in the west (including Northrop Aviation Corporation, formerly Avion Corporation) became The Boeing Company, headquartered in Seattle. Company branches in the east became United Aircraft Corp with Pratt & Whitney, Vought Aircraft, Sikorsky Aircraft, and Hamilton Standard divisions. Further, United Airlines was to become free from any corporate connection with either engine or airframe manufacturers. William Boeing retired from the company at this time. He was very disappointed by the outcome of the antitrust battle with the US government. He sold his stock in UATC and/or The Boeing Company and left the office and aviation. He turned his attention to land development and raising horses. For the industry, one consequence of these events was that the airlines shifted focus from carrying mail to carrying passengers.

In August 1935, the US Army Air Corps (USAAC) posted requirements for a new bomber. The Boeing Company, now without its former leader, responded

Fig. 22: Top, A TWA Douglas DC-3B in flight. Note how the main landing is partially retracted. (The Museum of Flight Collection). Bottom, a Boeing model 307 Stratoliner in flight. Note propellers on right wing are feathered, and vertical tail is set to counteract the imbalance. (Boeing Images BI232860).

with a four-engine Model 299 that would become the B-17 while Douglas studied and proposed a twin-engine B-18. The more capable four-engine B-17 had enthusiastic supporters in the USAAC but it was more costly than the B-18. A year later, the contract was awarded to Douglas, presumably for cost reasons. Interest in the B-17 remained high within the USAAC however, and monies were found to fund continuing studies and construction of a number of YB-17 prototypes. For this work, Boeing ultimately landed a contract to build 512 of the B-17s in 1940. The War Department in the United States was beginning to see the clouds of war on the horizon.

To stay in the airliner business and recover from the poor sales performance of the 247, Boeing used the work that went into the B-17 to shape a civilian airliner. This became the Model 307 Stratoliner, which first flew in 1938 and was the first airliner to feature a pressurized cabin. Four 1,100 HP Wright GR-1820s,[7] that also powered some versions of the DC-3, propelled the 307. The power available at this time is a tribute to the progress made by the engine-builders.

Cabin Pressure

Pressurizing the cabin has an important effect on the design of the fuselage cross-section that previously was often rectangular. The new circular-cylindrical fuselage shape allows the stress due to the inner higher pressure to be taken up efficiently by the fuselage skin and the ribs that determine the form, much like the hoops that hold a barrel's shape. Boeing was again a pioneer in incorporating new technology. It was, however, unable to capitalize on innovations to outdo its competitor, Douglas. The production of the 307 was limited to ten aeroplanes that were flown in service with Pan American and TWA.[8] A number of 307s served later as military transports.

The aeroplanes designed in the early 1930s had limitations that seriously affected the economics of operations. The aeroplanes of that time, principally the 247, the DC-2, the Ju52 and even the Lockheed Super Electra (1937), were small with 10 to 17 seats. The advent of the DC-3 (1936) changed the industry when that aeroplane offered a capacity of 32 seats. This aeroplane established Douglas Aircraft Company as the leading transport-aircraft maker. Boeing's Model 307 had a competitive capacity of 38 seats.

---

7. The Wright GR-1820 had the following performance indices: 0.46 HP/cu. in. and 1.2 lbs/HP. The 'G' in the number refers to the inclusion of reduction gearing on the output shaft.
8. Charles Lindbergh was associated with this airline. Howard Hughes would later control its destiny.

Fig. 23: Left: Museum display of the round fuselage shape required by a pressurized interior. This is a section of an Airbus A-300 (wikipedia). Right is a view of a Model 377 Stratocruiser in the manufacturing process. The floor panel is designed to withstand the tension exerted by the outward pressure on the two portions of the fuselage. (Boeing Images BI228135).

For all manufacturers, the start of the Second World War unfortunately prevented the passenger market for business or pleasure air-travel from thriving. On the positive side for the manufacturers, however, these same aircraft would play an important logistical role in the conflict. More importantly, military funding would be available for the achievement of improved performance. The Douglas DC-3 was produced as a civilian aeroplane until 1942. The military DC-3 derivatives, the C-47 and C-53, were built in numbers exceeding 10,000, eclipsing the civilian sales made earlier.

Octane and Doolittle

The noise of coming war had a substantial effect on the engines that powered aeroplanes. The gasoline used by aeroplanes in the 1930s was 87 octane aviation-gasoline, or 'avgas'. Major James H. Doolittle of the USAAC was concerned that if war were to come, the United Sates would need a higher octane fuel for better engine performance. He succeeded in getting the Shell Oil Company to produce a new avgas with an octane rating level of 100. He convinced the company to proceed with mass production and stockpiling it. By wartime in 1941, the amounts on hand were a good start for flying military aeroplanes with it. During the transition time to the higher-octane gasoline, engine manufacturers were also keen to exploit the availability of this new fuel. They increasingly specified avgas 100 for their engines because it allowed them to raise engine compression ratio. The fuel had then and still has the ingredient tetraethyl lead to combat detonation.[9] The use of high-octane gasoline in aircraft engines ultimately became almost universal, especially during the war on the Allied side. Lower octane fuel became motor gasoline or 'mogas'.

It would be an omission not to mention the work on instrument flying among Jimmy Doolittle's many contributions to aviation in the 1920s. It is hard to imagine today's airliner operations at night or in bad weather without the ability to fly using only instruments. In the mail contract days, navigation was sustained by markings on the ground and on buildings. To this day, large letters made of painted rocks on the sides of hills identifying the neighbouring town may be found as a reminder of the era. Later, during the war, Lieutenant Colonel Doolittle went on to acquire fame by directing the raid on Tokyo in 1942 in B-25s launched from an aircraft carrier.

Flying Boats

The 1930s spawned another innovation, this one designed to overcome the limited airport infrastructure in some areas of the world and the limited range of aeroplanes of the day. For example, Pan American Airways (PAA, also referred to later as Pan Am) and other airline operators wanted to serve the Caribbean and South American markets from points in the US. The cities in this region were readily reached by aeroplanes of modest range, if one could hop from one airport to another, much as a train travels to stations between the end points. The intermediate stopping points were required for refuelling and passenger exchanges, but they often lacked airports as such. A logical solution was to take advantage of the large bodies of water near many cities, using them to 'land' aeroplanes instead of conventional runways.

The obvious answer lay in designing an aeroplane able to take off and land in water. The limited range of aeroplanes at the time was crucial in focusing attention on the 'flying boat' solution. Among others, which included the Glenn L. Martin Company and the Consolidated Aircraft Company,[10] the aviation pioneer Igor Sikorsky saw the opportunity to design and build flying boats. Beginning about 1928, he designed and built a number of such aircraft that Pan American Airways

---

9. Detonation is the potentially damaging self-ignition of the engine's fuel mixture during combustion.
10. Consolidated had built a number of twin engine 'Commodore' flying boats for PAA. Consolidated ultimately became Convair after a 1943 merger with Vultee Aircraft Corporation.

and others used to serve specialized markets. The designs were designated S-38 to S-44, with some of the numbers in between built only as prototypes.

The S-38 was a twin-engine, semi-biplane flying boat that was built in relatively large numbers (101). The 10-passenger S-38s were used by PAA where Juan Trippe was beginning his career as the principal force behind that airline. Other operators also used the S-38. It featured 400 HP Pratt & Whitney R-1340 Wasp engines and was small enough to be privately owned by wealthy people including Howard Hughes and Charles Lindbergh. A smaller single-engine S-39 was also built but did not serve as an airliner.

A desire for greater passenger capacity and range on the part of PAA moved Sikorsky to build the S-40. It was introduced in 1931 as a larger version of the S-38 and the technology used in it was well in hand at the time. The S-40 monoplane could carry thirty-eight passengers and was powered by four Pratt & Whitney Hornet-9 engines. As a consultant to PAA, Charles Lindbergh found the S-40 to be old-fashioned in that it featured a heavy array of struts and wires to connect the wings to the fuselage and pontoons. He described it as a 'flying forest'! Only three S-40s were built. Lindbergh's evaluation caused Sikorsky to develop the S-42. It was similar in size but much cleaner aerodynamically and with modern features such as wing flaps and variable pitch propellers that had become available by this time (1934). PAA was the only customer for the S-42 that could not only accommodate 37 passengers in seats but 14 sleeping berths as an alternative. Sikorsky built 10 of these aircraft in three variants. The range of the S-42 was 1,930 miles (in practice typically 1,200 miles) and its cruise speed was around 150–160 mph.

A 'Baby Clipper', S-43, was a follow-on purchase by PAA. This twin-engine aircraft carried 18–25 passengers beginning in the mid-1930s. It was almost the last of the flying boats designed by Sikorsky.

The limited capacity of the S-42 and the good market conditions in the southern hemisphere as well as in the Pacific caused PAA in the mid-1930s to revisit the aeroplane design issue. It asked the Martin Company to build a higher performance clipper. Martin produced the M-130 flying boat with a capacity about the same as the S-42 except that it had more powerful engines: four Pratt & Whitney R-1830 Twin Wasps of ca. 900 HP each and longer range. The S-42 used 660 HP Pratt & Whitney R-1690s.

The issue for PAA was crossing the Pacific Ocean to reach the Asian market. The largest open water span was the 2,400 miles from San Francisco to Honolulu. The M-130 bridged this longest leg of the 8,000-mile island hopping trip and set the stage for commerce across the Pacific by air. The stops beyond Hawaii were on the islands of Midway, Wake, and Guam to reach Manila. From there,

China was within reach. These stopover islands had been prepared with facilities for refuelling as well as passenger and crew rest, and unfortunately went on to become battle sites for the war in the Pacific in the next decade.

The first of the three M-130s built was named by PAA as the *China Clipper*. In the public's eye, this name was transferred to the fleet of flying boats on that route. The other two Martins were the *Philippine* and the *Hawaii* clippers. The other Sikorskys in the fleet had similar names reflecting the routes where they were used. Using the aircraft, PAA started airmail service in 1935 and passenger service the following year. The journey was long, almost sixty hours of flight time, and financially not within the reach of ordinary citizens during this portion of the Great Depression. It is estimated that the cost then ($799) was about twice the cost of a first class ticket today.

Business was good and projected to be better in the future and it was time to plan for it with a larger aeroplane. PAA approached The Boeing Company for a proposal that was to become the Model 314, a pressurized cabin flying boat design that could carry around 75 seated passengers (though about half that many when sleeping berths were used for overnight flying) over almost 3,700 miles at a speed of 188 mph. The longer range would allow the aeroplane to skip some of the stopovers put in place for the M-130s. The four-engine Boeing 314 *Clipper*

Fig. 24: Take-off by a Sikorsky S-42. (Igor I. Sikorsky Historical Archives © 2018).

was first flown in 1937 and presumably destined to become the PAA flagship for markets where its use was appropriate. When Boeing prepared its proposal, the 1,600 HP Wright R-2600, Twin Cyclone, had become available allowing for a significantly higher performance aeroplane.

In 1939, the clipper service was opened with scheduled transatlantic service by connecting Southampton (England) with Port Washington (New York). Three refuelling stops in New Brunswick, Newfoundland, and Ireland were required to make the crossing. It also operated in the Pacific. A trip on the Clipper was a luxurious and expensive affair with a correspondingly limited market appeal, much like that of the Concorde decades later. Unfortunately, as with the Model 307, the new world war hindered the Clipper's full development as an airliner. It was also difficult to fly and required special pilot training to handle the water operations. The cost associated with this issue and the limited market motivated the development of land-based aeroplanes with greater range. A total of twelve 314s were built.

Sikorsky attempted to maintain Pan American as a customer with a proposed large flying boat similar in capability to the Boeing 314. Three Sikorsky VS-44 aircraft were built and sold, but not to Pan American. The war and better land-based aircraft brought an end to the commercial flying boat era, not only for Boeing and Sikorsky but all the others as well.

An interesting aside is that the engines for the Boeing 314 with its four 1,600 HP Wright R-2600 Twin Cyclones were then state-of-the-art. The US was keen to

Fig. 25: A Pan Am Boeing 314 Clipper ready for departure. (Boeing Images BI228577).

Fig. 26: A cutaway Wright Twin Cyclone R-2600. The red edges were cut in the sectioning process. Note that fuel injection takes place (black carburettor) before compression by the supercharger. (Photo by author Museum of the United States Air Force, Dayton, Ohio).

keep them out of Japanese hands when the war in the Pacific boiled over. The story of recovering one of the Pan American aeroplanes from New Zealand, the *Pacific Clipper*, makes for good reading. It was successfully flown westward around much of the globe to avoid the Japanese navy in the Pacific Ocean.

The Boeing 314 ushered in the era of the big twin-row engines in commercial service. Both Pratt & Whitney and Wright produced engines of this type in the US. Similar engines were also produced elsewhere but they were the state-of-the-art in the late 1930s.

Most displays of engines are limited to front views. Below is a rear view of a Pratt & Whitney R-2800, a two-row radial engine showing the details that can only be seen there: engine mounts, intake and exhaust ports and/or manifolds, accessory access ports (hydraulic pump and electrical generator), etc.

Fig. 27: Rear view of a Pratt & Whitney R-2800 (Double Wasp) showing silver intake manifolds emanating from the outlet of the supercharger, red plugs where the exhaust manifold is connected, and engine mounts (four visible grey attachments at 11, 1, 3, and 5 o'clock positions). The carburettor is not present and is normally located above the black cover. Located at the Historic Flight Foundation in Everett, Washington.

The mid-1930s also saw the development of a uniquely British design for a two-row air-cooled radial engine. The Bristol Engine Company developed a family of engines that found limited application in the commercial world but were produced in large numbers for the military. The best known of that family was the Bristol Hercules, and its unique feature was the use of sleeve valves in lieu of the poppet valves that were the industry norm. The idea behind the development was a work-around to the problem of using four poppet valves on two-row engines. The sleeve valve design allowed improved compression ratio (or allowed use of lower octane fuel) and better airflow in and exhaust out of the cylinder. The large number of Hercules engines built seems to indicate that some of these benefits were realized. On the problematic side was the greater challenge of lubricating the large surface between the cylinder and the sleeve. A casual visitor examining a model of this engine will notice the absence of the usual pushrod tubes that adorn conventional radial engines and that the spark plugs reside on the top of the cylinder head. The sleeve valve was an interesting sideroad in the history of aircraft engines. It served its applications remarkably well with many variants produced from the late 1930s to the 1950s with power outputs between 1,300 and 1,700 HP.

An Aerodynamic Breakthrough

In 1935, at the Volta Congress in Rome, a German aerodynamicist named Busemann[11] presented a paper suggesting that the way to fly closer to sound speed was to sweep the wing. Forward sweep[12] and rearward sweep would both be effective, but today the rearward sweep is universally used in the design of airliners. The idea is that deleterious effects of wave drag are associated with the flow of air impacting the wing at *right angles* to its leading edge. When the wing is swept, flight speed could be higher than the air velocity component aimed at right angles at the leading edge. In effect, the wing appears to be thinner to the approaching air and that helps reduce drag.

There is evidence that the Hungarian-American aerodynamicist Theodore von Karman[13] attended this conference. He either did not note the swept wing idea or forgot it, so that its implementation in the US had to wait a while longer.

Not much attention was paid to the idea of wing-sweep except that it was noted by a number of German engineers who were designing military aircraft during the war. Although researchers were doing wind tunnel tests on swept-wing models with good success at the time, the concept did not find an effective application.[14] That would change in the post-war period.

---

11. Adolf Busemann worked in the German aerodynamics research community before and during the Second World War. After the war he emigrated to the US and joined NACA.
12. A Junkers Ju287 jet powered bomber prototype with forward swept wings was built and flown in 1944, but never put in production. In the mid-1980s, an American NASA X-29 was built with forward swept wings and tested for control and structural issues.
13. Director of the Guggenheim Aeronautical Laboratory at the California Institute of Technology after 1930. He was key in the creation of the Jet Propulsion Laboratory in 1944 and headed the American (aeronautical) fact finding team (called the 'Scientific Advisory Group') sent to Germany after the Second World War.
14. One could argue that the modest sweep of the Me-262 was employed to enable faster flight.

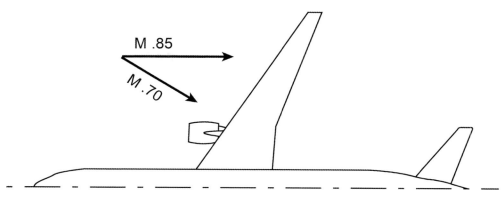

Fig. 28: The swept wing aeroplane. The sketch shows the wing, the long flight velocity vector (M .85) and the shorter velocity vector (M .70) normal to the wing chord that determines the drag rise associated with flight speed near the speed of sound.

Rigid Airships: the Road not taken

May 6th 1937 marked the end of a long history for a competitor to aeroplanes and ocean liners, the gas filled airship, also called a *dirigible* because it could be steered. The Zeppelin companies constructed airships of various kinds over many years starting before the First World War. Many of these dirigibles were used in that war. The Treaty of Versailles ending the war (1919) limited what Zeppelin could build. After a number of treaty amendments, they ultimately were allowed to build a number of large commercial passenger-carrying ships starting in 1928. These ships were slow (40–50 mph, 85 maximum) compared to airliners, albeit faster than ships, but they had the range to travel between continents.

A fiery hydrogen gas conflagration of the Zeppelin airship *Hindenburg* occurred on landing at Lakehurst, New Jersey, following a transatlantic trip from Friedrichshafen in Southern Germany. A thunderstorm in the area during landing was reported to be a major cause of the accident. The lack of access to helium as a fill gas was because the US refusal to sell helium forced Zeppelin to use hydrogen, an unfortunate decision that did not help promote this mode of transportation. The Zeppelin is, however, rich with history in war and peace.

The *Hindenburg* (LZ 129) was powered by four liquid-cooled Daimler Benz, rather heavy (over 3 lbs/HP), V-16 Diesel engines producing 1,200 HP each. The airship carried 50–72 passengers with an almost equally large crew of 40–60.

A new Aeroplane Engine

In the 1930s it became increasingly understood by engineers that the basic science for a new type of engine showed promise for faster aircraft propulsion. The fundamental idea was to devise an engine that could take in a larger amount of air. More air, more fuel … more power!

All engines require compression and the traditional way of doing it with a piston within a cylinder was fundamentally limiting. Thus arose the notion of a separate compressor processing air steadily. This naturally involved combustion in an equally steady manner, resulting in an altogether new engine cycle. These are the basic ideas of the gas turbine engine, the turbine being the component that is required to supply power to the compressor.

The compressor working to produce outlet air pressure sufficiently high was the design challenge. There are two ways of building a so-called aerodynamic or dynamic device for the compression of a gas. The two approaches involved very different types of machines. The easiest is the so-called *radial* flow compressor. In such a compressor, the pressure of the output air is raised by hurling the air radially outwards using an impeller (hence the name). The fundamental limitation of the radial flow compressor is that the outflow airspeed should, for reasonably efficient performance, be limited to less than the speed of sound.[15] A diffuser then collects this rapid flow and gently slows it down to recover the pressure. The speed limitation on the impeller or rotor's outer radius, the so-called *tip speed*, makes it very difficult to achieve pressure ratios greater than about three. Additionally, the rotor presents the flow with a fairly large surface area that induces rotational flow, all of which keeps the efficiency of the compression process from being little more than just adequate. The beauty of the radial flow compressor is its tolerance for varying flow conditions. In other words, it does not work all that well, but it does work reliably.

A more challenging way of compressing the air is by means of an *axial* flow compressor consisting of rotating blades that accelerate the flow by a small amount. An increase in the velocity component in the tangential or azimuthal direction is forced by the shape of the rotor blades. The air then enters a row of stationary blades (stators) fixed to the case of the compressor to slow the flow back down, restoring the original flow orientation and increasing the pressure. This has

---

15. The reason is that supersonic flow is notoriously difficult to slow back down to subsonic speeds, because shock waves are involved. For supersonic aeroplanes, Concorde for example, the slowing of air into the engine required a very complicated inlet to keep the losses in check.

advantages and disadvantages. The dominant characteristic of such a compressor arrangement is that the pressure rise is small. However, a number of such stages (consisting of rotor and stator combinations) can be arranged in a series so that any pressure ratio can be realized. That is an advantage. A distinct disadvantage is that the flow past such blading is required to be pretty well aligned with the blade geometry and failure to do so can result in a stall over the blade flow. Such a blade stall can cause the entire compressor to stall rendering it useless and the engine non-functional. Another characteristic of the axial flow compressor is that the surface area interacting with the flow is manageable so that efficiency is potentially quite good.

These were the issues confronted by engineers devising an engine. Two individuals who played pioneering roles in the development of a gas turbine for aircraft propulsion were Frank Whittle in Great Britain and Hans von Ohain in Germany. It is not knowable whether they were aware of each other's work but the historical consensus is that they worked independently. They both took advantage of what was known in their day, work done by others.

The idea of a jet propulsion engine was patented in 1930 by Whittle,[16] despite a negative review by A.A. Griffith commissioned by the British Air Ministry to determine whether the engine was of interest for military aviation. Griffith expressed the thought that the jet engine concept might not be suitable for aircraft propulsion. Griffith's credentials included authorship of a paper published in 1926: 'An Aerodynamic Theory of Turbine Design', in which he suggested the idea of an axial flow compressor and a turboprop engine.

In spite of this assessment, Whittle's company, Power Jets Ltd, proceeded to build such an engine in the following years. It is to his great credit that he carried on without much funding and support from industrial or government interests. Both were sceptical. The primitive means applied to the necessary experimental work that Whittle was forced to do seemed to reinforce their views of the impracticality of the jet engine concept. No small part in that scepticism was the notion that the established way of doing things had to be protected. Also to his credit is the challenge he overcame to tackle the combination of technologies that had to work together to have a viable engine: physical and functional compressor construction, compact and reliable combustion, turbine aerodynamics, engine controls, metallurgy, heat management, to mention just a few. The endeavour to

---

16. John Golley on p. 36 of his book *Jet* states that Whittle applied for a turbojet patent (UK no. 347,206) on 16 January 1930 that was (openly) granted in April 1931.

build his engine was not unlike that faced by the Wright brothers as they tackled the aeroplane as a system with its own multiplicity of challenges.

Simultaneously in Germany, Hans von Ohain had the same idea. Von Ohain studied under Professor Prandtl at the University of Göttingen. His studies in the mechanics of fluid flow led to ideas for a jet engine. A patent[17] was filed and recognized in 1935 and subsequently labelled 'secret' by the German government. This patent ultimately led to his meeting with aeroplane builder Ernst Heinkel who was taken with the idea of a jet-powered aeroplane. This encounter led to von Ohain's good fortune to be able to harness industrial support and expertise to build a demonstration engine. Thus he was able to overcome, rather quickly compared to Whittle, the hurdles of the technology.

Both of these young men, von Ohain and Whittle, used the radial flow compressor in their engines. However, their approaches to the turbine differed. Whittle used an axial flow turbine while von Ohain used a radial flow turbine. Von Ohain was probably the last to use the radial turbine in an aircraft engine application because the superior virtues of an axial turbine were readily apparent to everyone.

Historical records show that the first flight of a jet powered aeroplane was on 27 August 1939. The aeroplane was a Heinkel He-178 built specifically for a demonstration of the jet engine concept. Heinkel built not only the aeroplane but made his facilities available to von Ohain to build the engine. The historic aeroplane used in the first demonstration of jet-powered flight was lost during the Second World War bombing in Germany.

Whittle also worked on his engine and succeeded in a demonstration flight in a Gloster E.28/39 in 1941 with an engine (W.1A) producing 1,450 lbs of thrust. Whittle's road to this point in the evolution of the jet engine is an interesting story. He was a talented and driven man. The Battle of Britain in the summer of 1940 and the dire war situation in that country prevented Whittle from receiving as much government financial support as he needed to proceed more rapidly. During the Battle of Britain, the relationship between Great Britain and the United States, while not yet a military alliance, was close. In particular, they shared materiel and information to a great degree after the passage of the Lend-Lease Act in March 1941 when the US was still unwilling to enter the conflict. Importantly for our story, Great Britain passed the gas turbine technology to the United States and development progress was made rapidly in both countries thereafter.

---

17. German Patent 317/38 'Verfahren zum Umwandeln von Wärmeenergie zu Kinetische Energie eines Gasstromes' (Process for the conversion of thermal energy to kinetic energy of a gas stream), awarded 10 November 1935.

Jet Engine Challenges

Whittle's reality was that he needed to tackle three aspects of the engine. These were, starting with the easiest, the turbine, the compressor, and the combustor. The technology of the *axial* turbine was well in hand in Whittle's time. It originated with the steam power industry and continues to the present day in all jet engines. The second dimension, the radial compressor, was pretty well understood. Hence, he relied on it, rather than the axial flow compressor. As we shall see, future developments will reveal that the radial flow compressor as a part of the jet engine will have been a sideroad in the development of these engines. The use of the axial flow compressor would ultimately dominate the industry. Finally, the development of the combustor proved to be a significant challenge.

Whittle's biography is replete with the effort he made to realize a combustion chamber design that allowed the release of the required amount of heat in a volume that was sufficiently compact. Heat release rates were demanded that exceeded the norms of the day. A look at early British gas turbine engines will reveal the very large combustion chambers that Whittle used in the first practical engines. It is a marvel to consider the progress that has been made over the years to reduce combustion chamber volume in modern engines to the point where they are difficult to discern as part of the engine. The heat release rate per unit volume is about one hundred times greater today than it was in Whittle's day.

A challenge for the turbine was the heat to which the blades and the disc that held them in place were exposed. Without access to modern metal-alloys and air-cooling techniques, Whittle resorted to water-cooling in parts of his W.1 engine. That choice seems quite dramatic, questionably effective, and impractical for a real engine. But he had to start somewhere.

Regardless of turbine geometry, radial or axial, the function of the turbine is to provide the power to run the compressor. The power is taken out of the hot flow by the turbine as the gas (actually a mixture of air and combustion gas) pressure falls. By virtue of the heat added to the engine in the combustion chamber, the pressure in the turbine exhaust is still high enough after the power removal by the turbine to create a jet by means of yet another nozzle.[18] This jet and its momentum produce the thrust to drive the aeroplane. Here is the central principle of the jet engine. It can be built with axial and with radial flow components. The latter was initially easier, but the former was far better for many reasons.

---

18. Why the engine functions to produce thrust is explained in the Appendix. The explanation is in terms of internal combustion engine thinking with which the reader may be familiar.

In roughly the same time period, researchers at the University of Göttingen[19] worked on air compressors and succeeded in determining that a multi-stage compressor with aerofoils on a rotating shaft was possible and practical for an aeroplane engine. The flow through such a compressor is primarily along the direction of the rotating shaft and is therefore called an *axial flow* compressor. Such a compressor is shown in the cross-section sketch of the Jumo 004 (see Fig. 30, page 79). In actuality, the flow is helical and must be straightened to leave the compressor in the axial direction. How such a compressor works is left to the reader at this point because the description involves flow-velocity diagrams and mathematics to bring out its characteristics in full detail.

From the very beginning Whittle used the double entry radial flow compressor as a component in his engine designs, although there were voices in the British technical establishment expressing opinions that the axial-flow compressor might be better. Specifically, A.A. Griffith appears to have had second thoughts about the utility of the jet engine. His subsequent work led to the construction and testing of an axial flow compressor jet engine called the Metrovick F.2 by the Metropolitan-Vickers Company, a Manchester-based producer of steam turbomachinery. The engine ran successfully producing about 2,100 lbs of thrust and flew in a Gloster Meteor in November of 1943. It was judged too complicated for realistic use by the military and was consequently not put into production. In addition to heat management issues, the complexity had, in all likelihood, to do with the lack of an engine control system to help the pilot avoid compressor stall. That ended a skirmish in the battle between proponents of the radial and axial flow compressors in Britain, but only for a time. The radial compressor was more robust in dealing with compressor stall and thus won the day. That turn of events established Whittle, Rolls-Royce, and later General Electric in the US, as the driving forces that determined the near-term future of the jet engine with radial flow compressors in those countries.

In 1939 Griffith had joined Rolls-Royce and was an active and eventually successful proponent of the axial flow compressor within the company. Looking ahead briefly to the post-war period we note that development of the Metrovick engines continued to the point where the technology found its way into the Armstrong-Siddeley *Sapphire* engine. That engine was successfully used in a number of British military aircraft and was licensed to Curtiss-Wright as the J65. By the time of the Sapphire engine, Rolls-Royce would have switched to the axial flow compressor in its Avon engine, essentially ending the arguments between radial and axial flow compressors.

---

19. Where von Karman studied under another famous aerodynamicist, Ludwig Prandtl.

Everything Changes!

The Aerosphere collection (see bibliography) of engine and aeroplane data in 1939 paints a fair picture of the aeronautical world of that day. It is a richly illustrated, detailed, and an essentially complete description of the industry at that time. It would be overwhelming to list the number of engine and aeroplane builders who were part of the industry. In addition to the hundreds listed for the US, Germany, France, and England, there were many noted for smaller countries as well. Most of these manufacturing enterprises vanished after the war.

The compendium also notes aeroplanes that were attempts to be part of the 'coming' air transport business. There were British flying boats from Short and from Saunders Roe, a land-based airliner named the de Havilland DH-95, and an Armstrong-Whitworth airliner with a capacity of 27–40 passengers. Junkers in Germany had models Ju86 and Ju186 and there were types built by Dornier and by Focke-Wulf. The destiny of these aeroplanes would largely be determined by the war and few were part of the story of airliner evolution.

American air transport models from Lockheed include the L-12 and L-18 as variations on the L-10 built earlier. Lockheed had a Model 44 (named Excalibur) in the works that ultimately became the Model 049. Its history involved the manipulative input from Howard Hughes who, by that time, owned much of TWA (Transcontinental and Western). The Model 44 (later called the 044) was to compete with the Boeing 307 and Douglas aircraft.

December 7th, 1941. The United States entered the war when Germany declared war on it after the Japanese attack on Pearl Harbor. All engine developments thereby centred on the war and planning for peaceful commercial exploits evaporated. The Aerosphere ceased publication during the coming war years. It was a new time.

Chapter 4

Engines for the War Effort

The Second World War provided a great impetus for developing aeroplanes to achieve greater capabilities. The most important avenue for performance improvement was increasing the power output of engines. This was the necessary means to allow building heavier bombers and transports and faster fighter aircraft. For example, the B-17 was one of the heavy bombers during the early part of the war using Wright R-1820 engines of 1,200 horsepower. The B-17 had a maximum gross weight of about 65,000 lbs. The newer B-29, put in service in May 1944, could take off with twice the weight. This was made possible by the availability of the turbocharged Curtiss-Wright Duplex Cyclone (R-3350-23) engine with 2,200 HP, roughly twice the power of the B-17 engines.

In that period, the competitive and similar Pratt & Whitney Double Wasp R-2800 (see Fig. 27, page 63) was ultimately rated at similar power levels and used in airliners like the DC-6 and the Martin 4-0-4. Part of the power increase was obtained from water injection into the cylinders to allow the air mass processed to be greater and thereby produce more power. After the war, the Duplex Cyclone found application in the DC-7, the Lockheed Constellation, and Super Constellation.

By the end of the piston-engine era in the 1950s, the maximum power available from an engine (Pratt & Whitney R-4360) was over 4,000 HP, though it was typically rated between 3,500 and 3,800 HP. The R-4360 was a supercharged engine constructed of four rows of seven cylinders. It powered the last of the large piston-engine powered aircraft, military as well as airliners. These engines powered the B-29 replacement, the B-50, the Boeing Stratocruiser (Model 377), the Convair B-36, and, interestingly, the Hughes H-4 Hercules, the *Spruce Goose*.

As the reciprocating engine era comes to a close, we note some of the important parameters that describe them as they evolved from the 1920s to their last days in the late 1950s by noting specific engines' characteristics. The examples shown in the table below are representative and even they came in a number of versions. They generally grew in power output as they were developed. The power quoted is often maximum take-off power and that level is used only for a limited time period. Over the years, engines improved in ways important to the aeroplane designer. These ways include the amount of power from a unit of displacement and the engine weight per horsepower. These parameters describe engine size and weight for the power produced. They are quoted in terms of HP/cubic inch displacement and in lbs/HP to describe engine weight.

While displacement is commonly given in various units of measure, we note litres here to describe engine size. Cubic inches were standard in the early days of the engine business but nowadays the metric litre is well appreciated. Since the technology of internal combustion engines ultimately led to a power output of about one horsepower per cubic inch displacement, the engine size in cubic inches is a fair, albeit approximate, measure of power output. Rather than quote years to pinpoint engine technology, we note examples of the aeroplanes on which the engines were used. All the engines are air-cooled radials except for the Gnôme and two liquid-cooled V-12s.

Typical applications	cyl.	Litres	HP	HP/cu in	lbs/HP	Ref
1903 Wright Flyer	4 in line	3.3	12	0.06	15	
WW I military pursuit	9 Rotary	59.1	165	0.17	1.76	1
TBC 40 (Liberty)	V-12	27	400	0.24	2.11	
TBC 247	9	22	600	0.45	1.54	2
DC-2, B-17, TBC 307	9	30	730	0.46	1.19	3
DC-3, B-24	2x7	30	1500	0.66	1.04	4
B-50, B-36, TBC 377	4x7	71.5	4300	0.99	0.91	5
military fighters	V-12	28	1500	0.88	0.95	

Table 1: The evolution of air-cooled reciprocating engine performance together with two examples of liquid-cooled V-12 engines for comparison: Liberty and Allison V-1710, used in military fighters during WWII. The performance is given for 1. Gnôme Monosoupape N (Ref. Angle, G.D.), 2. Pratt & Whitney R-1340-S1H1-G, 3. Wright Cyclone GR-1820-F53, 4. Pratt & Whitney Twin Wasp R-1830-S1C-G, 5. Pratt & Whitney Wasp Major R-4360. TBC refers to The Boeing Company's model designations. To illustrate the lack of 'firmness' in these numbers, we note that over the life of these engines, the Pratt & Whitney Twin Wasp was rated from 800–1300 HP and the Wright Cyclone was designed as models ranging from 600-1300 HP. The rated power for the R-4360 was reduced substantially for use in long range cruise.

In three decades of development history from the mid-1920s to the 1950s, reciprocating engines doubled in horsepower per unit displacement and became significantly lighter. The weight difference between liquid-cooled and air-cooled engines in the last stages of their development history seems minimal.

During the peak year of the war, 1943–44, about a quarter of a million engines were produced for aircraft of all kinds. About one out of six were liquid-cooled and the remainder were air-cooled. The Wright (9-cylinder) R-1820s powering

Fig. 29: P&W R-4360 cutaway. Note four rows of cylinders and the radial flow supercharger at the rear end (at left) (Photo by author at the Museum of Flight). Bottom: Viewed from the front showing cooling air entry (Photo by author at New England Air Museum).

the over 12,000 B-17 Flying Fortresses and the Pratt & Whitney (14-cylinder) R-1830 powering the more than 18,000 B-24 Liberators dominated the wartime production picture, in part because the aeroplanes had four engines each.

It is interesting to note that in the last phases of the piston-engine era, a number of attempts were made to better the performance of these engines. During the war, water injection, specifically injection of alcohol-water mixtures, was implemented but only for fighter aircraft engines to give the pilots the possibility of applying large, momentary boosts of power. Generally, bomber and transport plane engines did not adopt this enhancement.

Pratt & Whitney was a reputable builder of the modern air-cooled engines but that does not mean that the company did not investigate alternative engine designs, specifically water-cooled engines. They found that these engines, in various configurations, did not offer advantages over the well-functioning air-cooled designs. Even sleeve valves were tried with the support of the engineers at Bristol. Pratt & Whitney chose to focus on what they knew and did well.

As the R-4360 seemed to be as large an engine as could be practically built, performance enhancements were attempted when the engine needed to be examined for possibilities of greater output. Connors, in his story of Pratt & Whitney, notes that compound-, turbo-supercharged, and jet thrust versions were contemplated.

One interesting avenue of pursuit might have opened the door to greater power. Consider that during the war, Wright Aeronautical built an engine that might have become a model for a later version that surpassed the R-4360 in many ways. The R-2160 (Tornado) was a six-row, seven-cylinder radial engine, *i.e.*, 42 cylinders in all. It had to be water-cooled because air-cooling tested the limits of that approach. It was installed in a number of experimental military aircraft but never produced in series. It performed rather well with a specific power of 1.08 HP per cubic inch and a weight of 1.03 lbs/HP.

The battle between proponents of air and liquid cooling of piston engines for large commercial and military aircraft was largely resolved with the abandonment of both approaches because of the limited potential for higher power. Time was on the side of the gas turbine and options to improve piston engine technology for high power were largely abandoned.

We again jump out of our time-narrative here to look ahead to the ultimate destiny of the large reciprocating aircraft engine. We note the characteristics of the gas turbine[1] and the ways it can and will be modified for purposes other than producing a high-speed propulsive jet. The story will eventually bring out the

---

1. The name 'gas turbine' is a more general description than the 'jet engine' because it includes the many variations that can be built around the basic components.

notion of the fanjet. Further, *all* the power inherent in the jet can be removed by a turbine to produce shaft power. Typically, the turbine executing this task is a 'free' turbine, connected only to the propeller as the load and free from a mechanical connection to the basic engine that is often referred to as a 'gas generator'. That configuration makes the engine functionally similar to a piston engine.[2] With the free turbine shaft driving a propeller, the propulsion system becomes a *turboprop*, that is in wide use in aviation today. The demise of large reciprocating engines can be summarized with the following postscript, centred on the availability of turboprop engines.

In the 1960s, the engine manufacturer Lycoming produced an engine destined for medium sized military aircraft, including helicopters. The T55 produced about 4,000 shaft horsepower. It was about as powerful as an R-4360, and the key point is that it weighed in at only 0.16 lbs/HP. A turboprop would be *one sixth* of the weight of a good reciprocating engine. In the face of competition from a simpler and lighter, albeit thirstier, engine, the aircraft designer's decision on choice of engines is relatively simple. The Lycoming engine is but one of a number of engines in that size class, representing the state-of-the-art at that time.

Gas turbine engines are costlier than piston engines so that the latter are still used today in small aircraft where cost is critical. Aeroplanes using these engines are not flown around the clock and year as an airliner has to be. Rather, such aeroplanes tend to be used relatively infrequently as standby tools for business or, often, for pleasure. Piston-engines in small private aircraft differ from large reciprocating radials in that they are mostly built in 4- or 6-cylinder horizontally opposed configurations. For example, a Continental O-470 ('O' for horizontally opposed cylinders and 470 cubic inch displacement), a popular six-cylinder engine for small civil aviation aircraft like the Cessna 180, produces about 240 HP. The descriptive numbers from sales brochures can vary for different configurations. They suggest, however, that such a modern engine can be characterized by 1.6 lbs/HP and 0.5 HP per cubic inch. When one compares these approximate numbers to the performance of the last generation of large high-power engines, the degree to which conservatism exercised for the sake of long life and reliability is evident.

Icing

The 1930s and the Second World War experience led to a number of innovations that later found application in the design of later commercial aircraft. One of these

---

2. The formal name of the engine with shaft power output is *turboshaft*.

was the development of de-icing of the leading edges of wings and propellers. Encounters with flying conditions where ice formation was an issue brought out ideas about defeating the dangerous situation of icing that could easily destroy the lifting capability of a wing or propeller blade. The inflatable rubber leading edge of a wing that could remove ice from the wing by a slight inflation and shape-change was a boon when deployed in 1943. Propeller blades were typically de-iced with electric heating elements in the blades.

Today's aircraft bleed air from the jet engine compressor to the wing leading edge to the same effect. Even direct application of de-icing fluid through many small holes in the leading edges of wings is used on modern aircraft.

Tricycles

The war years marked the end of the use of tail wheels in favour of tricycle landing gear. The tail wheel was often necessary when ground clearance concerns for the propeller were an issue. On the military side, the Consolidated B-24 that first flew in late 1939 had tricycle landing gear. On the transport side, the DC-4 featured this arrangement beginning in 1942 as did the new B-29 being designed and the later Boeing 377. In contrast to the tail wheel, the tricycle gear allowed for a level cabin and provided a better view for the pilot when the aeroplane was on the ground.

New Technology in Aircraft Propulsion

The notion of a jet aeroplane can be said to have been tried with a piston engine driving a propeller shrouded by the aeroplane body. Such a configuration was tried by the Italian firm Camproni. The aeroplane, the Camproni Campini N.1, was built as a prototype and flown in 1940, but it would be incorrect to say that it was a 'jet' aeroplane. While it looked like a later genuine jet aeroplane, it was a piston engine aeroplane creating a propulsion jet, much like any propeller-driven aeroplane engine creates a jet for thrust. The difference is simply that the jet pressure is created in an internal propeller and the pressure driving the jet velocity was slightly higher than that of an ordinary propeller. It was an interesting but unrealistic idea. Dreams of evolution into a military aircraft were dashed by the performance: it was short of power.

In Germany, as in Great Britain and the United States, the stage was being set to begin the jet age, driven by gas turbine engines. The interesting and complicated story of the early German development towards a practical engine may be found in an article written by D. Culy and published by the Aircraft Engine Historical

Society. In short, the research work at the University of Göttingen regarding the feasibility of an axial-flow compressor signalled an opportunity to exploit that technology to build what is a 'jet engine' to propel an aeroplane. In July 1939, the Reich Luft Ministerium (RLM, or Reich Air Ministry) awarded contracts to the firms Junkers and BMW to design and build axial-flow jet engines. There were other manufacturers in Germany with ongoing interests in the jet engine, including Heinkel and Daimler Benz. Heinkel continued the collaboration with Hans von Ohain that started in 1936 and led to the first flight of a jet aeroplane.

A number of configuration variations were considered for development, but limited resources and political options allowed only the rather simple turbojet for aircraft propulsion to be funded. Among the types of engines considered were what we call *bypass* and *turboprop* engines today. The bypass concept was actively investigated by Daimler Benz in an engine called the DB007. It was not likely to have been supported by the Reich Air Ministry and certainly did not reach a satisfactory state by war's end. Daimler also investigated an interesting compressor configuration where counter-rotating 'stator' blades were used in place of the now common stationary stators.

The Junkers and BMW efforts were aimed at construction of the Jumo 004 and the BMW 003. The latter was to be a higher performance engine but it was not completed as a production engine before the war ended. The leaders of these two efforts were Anselm Franz at Junkers and Hermann Oestrich at BMW. The engines resulting from development efforts were considered by aircraft manufacturers for various types of military aircraft, a number of which flew as prototypes, and two types were mass-produced.

Jumping ahead to the time after the war, we note that Franz went on to the United States after the war and became vice-president of the AVCO Corporation as chief of the Lycoming Gas Turbine Division and Oestrich went to France to head the technical effort in what was to become the French gas turbine engine company SNECMA (later Safran). The author's father started his engineering career at Junkers working under Franz and, after the war, followed opportunities first in France under Oestrich and then in the United States, working with Franz a second time.

Critical Engine Control

The designers of the Jumo 004 understood that it was unrealistic for the pilot of an aeroplane powered by such an engine to deal with the requirements of the compressor. These were to avoid conditions that lead to stall and stay near the

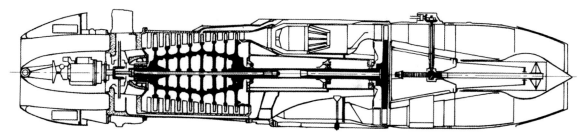

Fig. 30: Cross section of the Jumo 004 engine. Note the simplicity of the design and its streamlined shape. Eight compressor stages are used and a single turbine stage. The blades on the compressor discs on the central shaft are rotor blades and those affixed to the stationary case are stator blades. Note also the diffuser between the compressor and the burner to reduce the air speed ahead of fuel addition. In the nose cone of the engine is a small pull-started two-cycle gasoline starter engine.

point of most efficient operation. Of these, the stall problem is most serious.[3] Because the compressor blades are like small aerofoils, they can stall like a wing that loses lift at high angles of attack. This event renders the engine as a whole non-functioning. Besides leaving the aeroplane without power, compressor stall can be visually dramatic, especially in modern engines with high compression-ratio compressors. The functional failure of the compressor allows the high-pressure fuel-air mixture in the combustor to leave the engine in the forward direction where it will appear as a short-lived, spectacular (but almost harmless) fireball.

In practice, early attempts at flying the Jumo 004 proved full of difficulties associated with control of the engine. Gauges could have been installed to inform the pilot of the engine flow conditions, but flying the aeroplane, especially in a wartime setting, is complicated enough. The engine had to have a control system that shields the pilot from having to make any adjustments to the engine other than to demand more or less thrust. In the absence of such a control system, the pilot would have to 'fly' two systems: the aeroplane and the compressor.

The Jumo 004 was therefore designed to use the only option available at the time to manage the compressor airstream: the flow area of its main propulsion nozzle. This flow-area could be varied by moving a centrebody in or out of the convergent propulsion nozzle at the exhaust end of the engine. This body was humorously referred to as the 'onion' by the engineers.

---

3. We will not distinguish between complete stall and stall in a limited section of a compressor, which is often referred to as surge. Both are serious operational conditions that are best avoided.

Fig. 31: A museum display of the movable centre body on a Jumo 004 in a cutaway engine. The body touched on by the author's finger (the 'onion') is made to move fore and aft by an actuator sensing flow conditions. The air or, better, the mixture of air and combustion gas, from the partially visible turbine flows towards the observer. (Photo by Glen Ferguson at the German Museum of Technology [Deutsches Technikmuseum], Berlin).

The centrebody's position was fixed by a control system measuring shaft rotational RPM and pressures that could be interpreted as airflow rate through the compressor. By these means, the flow through the engine could be controlled to maintain good compressor function. Such a system was designed and patented (in Germany during the war and later in Europe and the US) by the author's father.[4]

On the Jumo 004 the control task was performed by a hydro-mechanical system with pistons moving actuators. Today control is no longer carried out with variable geometry at the propulsion exit nozzle (except when afterburning) but is instead achieved with controlled variation of the stationary blade angles in the compressor (see discussion of this innovation later in our story) and by the use of multi-spool

---

4. US Patent 2,688,841, 'Control Device for Gas Turbine Propulsion Plants', S.H. Decher and W. Stein, 14 September 1954.

engine components. Nowadays the problem of compressor stall is an uncommon concern to pilots, except perhaps when a goose tries to fly through an engine! Flow control is also managed in modern engines with bleed from the compressor, during start, for example. The associated bleed valves are closed during normal operation.

Development of Jumo 004A engines initially included use of critical materials, specifically metals for alloying turbine blade steel for high temperature tolerance: cobalt, molybdenum, and nickel. These prototype engines first ran in October 1940. By January 1942, a 10-hour run and a thrust level of 2,250 lbs were demonstrated. The first flight in a Me-262 was in May 1943. The turbine alloying metals used in the early prototypes were not available for the production version (004B) in the wartime setting. The engineers had to make do with active cooling of the nozzle blades (see Fig. 32, page 86) with compressor air and coatings on the turbine blades to help resist the hot combustion gas exposure. Nevertheless, the B engine produced 2,000 lbs of thrust, albeit with a short running time of roughly 10 to 20 hours, sometimes less, between overhauls to repair or replace sections of the engine exposed to high temperatures.

The control system in the Jumo 004 engine had to be developed to make its operation practical. On the other hand, Whittle was fortunate in that the radial flow compressor is less prone to stall; in a way, the nature of the radial compressor obviated the need for a sophisticated control system. Unfortunately, the radial compressor type is also limited to modest pressure ratios where higher values were needed for better performance. In part, that limitation ultimately led to an almost universal use of the axial compressor in jet engines after the war.

Combustion in the Engine – the Easy Part?

In the foregoing parts of our story we focused on the mechanical components of the jet engine, the compressor and the turbine. Just as necessary is the processing of fuel for heat. That is not straightforward in the gas turbine engine as the Whittle story illustrates. Nevertheless, it was quickly and successfully developed to a high degree of functional perfection.

In the piston engine, fuel in the form of volatile gasoline is added to the air by means of a carburettor or fuel-injector and the air-fuel mixture is compressed and ignited by a spark. That sounds easy and it is, relatively speaking, because all that is required is that the mixture be close to perfect in the sense that just enough fuel is added to burn all the oxygen out of the air. The important performance dimension was and is that combustion takes place reliably. From an engine design viewpoint, the compression ratio had to be limited to values so that explosive ignition did not take place before the piston reached top-dead-centre. Since the time of Nicholas Otto[5] and the invention of the internal combustion engine (the ICE), that aspect of performance has been improved to a high degree. In the early days of aviation, piston engine technology was primitive to be sure, but it was good enough to work for the Wright brothers and almost all who followed. The industry succeeded with the reliability issue with innovations like dual ignition, *i.e.*, two independent spark plugs per cylinder and independent electrical systems to run them.

Combustion takes place in the gas turbine engine as a steady flow process. This is in contrast to conditions in the ICE where combustion takes place in a few thousandths of a second and results in very large increases in *pressure* to act on the piston. In the gas turbine, it is carried out in a combustion chamber or burner where the pressure stays fairly uniform with a steady air inflow from the compressor. The burner's role is to mix a steady flow of fuel and air, and do it efficiently, steadily, reliably and, more recently, in such way as to minimize pollutant production. In contrast to the piston engine, where pressure is increased by combustion, the result of this combustion process is to increase the *volume* of the flow. With a greater volume flow rate, the forces on turbine blades are greater and the power from the turbine correspondingly greater than that required by the compressor. By these means a net power is realized, and that is the basic idea behind the engine as a power producer.

The difficulty in designing a satisfactory combustor is that airflow speed out of the compressor is high. Even if one could mix the fuel and air ahead of the region of combustion, the flammable mixture has a limited flame propagation speed. That speed depends on the combustion conditions, but is relatively slow. A flame started with an igniter is easily blown out unless special steps are taken.

Gaseous fuels, like methane or hydrogen, have higher flame propagation speeds but are not of interest in aviation because storing them on an aeroplane involves heavy tanks that take up a lot of space. When the first experiments with the gas turbine were initiated by Whittle and by von Ohain, gaseous fuels, specifically hydrogen, were indeed used just to get the experiments going. Eventually they switched to more realistic liquid fuels. So how did they do it? They did it in a couple of steps that are incorporated in burner design today.

---

5. Otto patented (1864) the first atmospheric gas engine, the precursor to today's internal combustion engine. That engine executes the thermodynamic cycle that bears Otto's name. It was further developed into the modern 4-cycle gasoline engine we know by Gottlieb Daimler and Wilhelm Maybach (1876).

The first step is to slow down the flow from the compressor with a device that increases the flow area called a *diffuser*. The cross section of the Jumo 004 (Fig. 30, page 79) shows the burner elements fairly clearly. A diffuser functions in the opposite sense from a nozzle and involves a flow area *increase*. A slow airflow is also required (for thermodynamic reasons) to keep the pressure in the burner as high as possible. After all, a lot of power went into making the compressed air and one would not want to lose any of the precious pressure.

The diffuser cannot realistically slow the flow so much that it matches any flame propagation speed, so the flow is made into a vortex which increases the gas residence time in the combustor. Only the amount of air necessary for a reliable combustion process for the injected fuel is treated this way and that represents a modest fraction of the total airflow. The rest bypasses the primary combustion zone for later dilution of the gas to a temperature level tolerated by the turbine.

Fuel is sprayed into this vortex to make a combustible mixture. With an ignition source at the time of engine-start, a flame is created as a burning vortex of mixed fuel and air. To insure that ignition is uninterrupted, a feedback mechanism has to be provided to bring gas that has just reached almost complete combustion in contact with a fresh unburned air-gas mixture. This is the function of the flow-field created, or hardware serving as a flame holder.

The resulting combustion gas is very hot and must be cooled by mixing the extra available compressor air. The cooler air is by no means 'cool' because, having undergone a compression of 20 or even 40 times atmospheric pressure (in modern engines), it is hot, but cooler than the combustion air! The dilution process often involves more burning of any fuel remnants in the primary burner to minimize the production of pollutants.

A characteristic of combustor design is the design of the sheet metal with holes and slots for the singular purpose of avoiding contact between the metal and the hot gas.

The Turbine

The compression section of the engine is central to this story, partly because it is relatively difficult to realize but the technology took years to develop. The turbine is somewhat easier but presents challenges of its own. Lest that part of the engine be neglected, the following few paragraphs are to shed light on the issues there and could be of interest to the reader.

The gas turbine engine relies on operation with hot gas; the hotter the better from the viewpoint of performance. The challenge is that the blades that orient the flow and the surfaces that contain it cannot stand the exposure to the gas temperatures one would like to have exit the combustor: sheet metal and the turbine blades would buckle, oxidize, burn through, or melt if one were to subject these materials to the combustion gas temperatures arising from burning fuel with just the right amount of air to remove all the oxygen from that gas. One refers to such combustion as 'stochiometrically' correct. The temperature of burning fuel with hot air when it has been substantially compressed reaches about 3,000F. The actual temperature level doesn't matter very much because we cannot use it.

The mechanics of the turbine have been understood since the 1920s as steam turbines and turbochargers were developed. In a gas turbine, the hot gas is first made to accelerate and turn by a stationary nozzle. This nozzle is a circular array of vanes similar to that in the compressor, but actually quite different because it functions in a reverse sense. In the turbine, power is taken *out of* the flow, while it is added *to* the flow in a compressor. From a gas dynamics viewpoint, the important difference between flow through compressor and turbine blades is that the pressure rises in the compressor's flow direction whereas it falls in the turbine. That fact allows the turbine blades to subject the flow direction to much greater angle changes than is allowable for the compressor blades. In turbine blading, the flow turning angles may be as large as 90 degrees while compressor blading angles are closer to 10 degrees.

The flow exiting the nozzle is directed against a turning rotor, and if the turbine has a single stage (see for example, a simple configuration like that of the Jumo 004), causes the flow to proceed to the propulsion nozzle with little or no rotation. If more stages are involved the process is repeated a corresponding number of times with the same final outcome.

Evidently, the turbine nozzle and the turbine rotor blades are subject to full exposure to the hot gas. This calls for a choice of materials that can take the heat without being structurally compromised. The spinning rotor also experiences centrifugal forces that are sizable and these must always be below levels that will cause a blade failure in tension. If that is to happen, and it was a concern in the early days of jet engine development, the outward spray of metal fragments is a serious safety concern. Of course, it can be contained with enough armour around the turbine, but that is heavy and unnecessary with good design and manufacturing. Examination of 1950s-era military jet aeroplanes highlight the risk with red stripes around the fuselage of an aeroplane or nacelle to warn service personnel to stay clear during engine run-ups, in case there should be a turbine disc or blade failure.

Beyond the choice of metal, the so-called superalloys, for the manufacturing of the turbine blade there is the means for casting the blades to realize a higher temperature capability. In the 1950s and '60s innovative casting technology was developed around directional solidification and single crystal castings for turbine blades. These methods avoid failure due to creep that occurs at metal grain boundaries by eliminating the boundaries altogether.

A good deal of pioneering work was done in connection with the metallurgical and manufacturing challenges associated with turbine blades. The literature is replete with reference to the various evolutionary steps but the name of F.L. VerSnyder[6] is associated with the birth of directional solidification. Further, B.J. Pierce is connected with single crystal development with a patent[7] application dating to 1965. Both men worked at United Technologies Corporation.

Material choice and manufacturing are, however, not sufficient if higher turbine operating temperatures are desired. Thus the universal approach to designing turbine blades (nozzle and rotor) is to include active cooling. Cooling air from the compressor is introduced into the root of the turbine blade. It circulates through the blade and finally exits the blade to join the primary flow. The cooling air exit could be at the blade tip or at the blade trailing edge. Such internal cooling is often augmented with external cooling introduced near the blade's leading edge by an array of small holes, and bathes the blade with relatively cool air. These techniques are in universal use in engines today and work very well. They have to, because cooling failure invariably leads to turbine failure; the turbine's metal cannot survive in the uncooled combustion gas environment.

The air used for cooling the turbine blades is hot and has value for the purpose of producing thrust by virtue of that heat and being incorporated in the jet stream. The mixing of the combustion gas and that air and the use of that mixture benefits the overall gas turbine engine performance in that all the available heat in the fuel is used for propulsion. Interestingly, this is in contrast to the reciprocation engine where a fraction of the fuel's heat is transferred to cooling fins or to a liquid coolant and irrevocably lost from the engine with no direct benefit for propulsion. The piston or Otto engine operates on a cycle that is at a fundamental level more efficient than the gas turbine cycle and can afford the cooling loss. Both engines have their limitations and yet both can be usefully employed in the service of propelling aircraft. This is the domain of engineers who wallow happily in thermodynamics!

Continuing progress in cooling, metallurgy, etc, has allowed effective gas temperature to rise over the years and, with it, engine performance. Naturally, there are additional issues that have to be dealt with in a good turbine design.

---

6. US Patent 3260505A dated 12 July 1966, 'Gas Turbine Element'.
7. US patent 3,494,709, dated 10 February 1970, 'Single Crystal Part'.

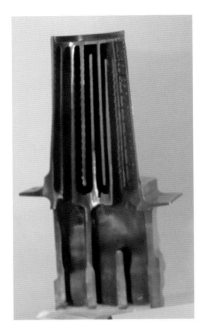

Fig. 32: At left: a first stage turbine rotor blades from two models of the GE CF-6 engine. The upper picture is of a blade cut open to reveal the tortuous cooling air path. The air is introduced through the root of the blade. The bottom left image shows the very small holes that are the exit points for cooling air on the blade surface. The film of cooling air shields the metal from direct contact with the combustion gas. Note the large turning angle of the flow through a blade row. To illustrate the path of cooling air (to the red stationary nozzle blades), the picture on the right is of the Jumo 004. The single turbine stage is shown in its entirety and the nozzle ahead of it partially cut away. The cooling air to the hollow nozzle blades is by means of the blue path at top right. The turbine rotor is uncooled. Hot gas paths are painted in red. (credits: CF-6 blades: Courtesy General Electric. Jumo-004 display: photo by author at the Museum of the United States Air Force, Dayton, Ohio).

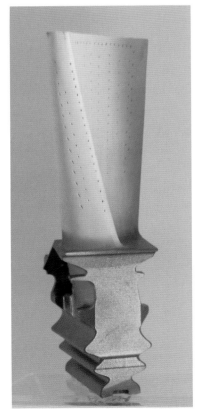

The materials have to be able to resist the strong oxidation environment and they must also operate to stay clear of destructive vibrations.

There are materials that can relatively easily meet many of the stringent design requirements of hot section parts of the engine at higher temperatures without cooling. These are the ceramics and ceramic-metal composite materials. They have been successfully introduced as coatings in the combustor and turbine nozzle but they have not,

Fig. 33: Another view of a modern turbine nozzle and its first stage rotor. One nozzle vane's cooling holes are visible (the other is the yellow covered one). From a sectioned CF-6 engine used for teaching purposes. (Photo by author at General Electric.)

heretofore, passed the test of benign failure on highly stressed components such as the first-stage turbine-rotors. In that environment, experience has shown that failure would typically be catastrophic, in the sense that it is sudden, because such materials tend to be brittle. Development work towards the creation of materials that combine the ductility and benign failure characteristics of metals with the temperature tolerance of ceramics is certainly underway. Such an achievement will result in a significant improvement in engine and aeroplane performance in the future.

New Aeroplanes – Luftwaffe Jets

Jet engines were tested on German prototype aeroplanes of various designs. None but the Messerschmitt Me-262 and one other aeroplane[8] with Jumo 004B engines entered service. Their entry was late in the war, towards the end of 1944. The Me-262 fighter was easily able to fly 100 mph faster than other fighter aircraft of the day … when the engines were in running order! That was a military advantage that the German air force, the Luftwaffe, hoped to use in the air-battle over Germany. Germany had lost air supremacy over its own country. The western Allies were advancing eastward from Normandy, westward from Russia, and northward from the French Mediterranean, and carrying out crippling bombing raids on German cities and military targets. The saga of the Me-262 includes Hitler's misguided notion to use the aeroplane not as a fighter but as a high-speed bomber. Hitler specifically forbade use of the aeroplane as a fighter for many months, during which Germany's drift towards losing the war became ever more inevitable. His thinking was that the expected invasion across the English Channel was going to require fast bombers to stall the invaders at the beach to allow German ground forces to reach the landing sites. The exact location of the landing sites was not known to the Germans in advance. Two aspects of the decision to use the Me-262 as a bomber were flawed. One was that the engines and the aeroplanes were not operationally ready, and the other was that accurate bombing was not possible with this aeroplane.

A couple of interesting books on aspects of the Me-262 in service are available, one by Luftwaffe General Adolf Galland, who wrote about the aeroplane from an operational viewpoint. The other is part of a book by Adam Makos, who tells a story of aviation heroism involving a Luftwaffe pilot interacting with a crippled B-17. The German pilot in this story was later involved in the use of the Me-262. Both are cited in the bibliography and make for interesting reading.

Hitler eventually retracted the order and allowed its use as a fighter, but it was too late for this new aeroplane type to play a pivotal role on Germany's behalf. Both sides did, however, recognize that a new age in aviation was at hand.

The air war over Germany in the summer of 1944 consisted of, in part, Allied bombing of fuel production facilities in Germany. By the middle of the summer, the production of fuel was so severely reduced that sustained air operations were not possible.[9] The end was near.

---

8. The twin engine Arado Ar234 bomber saw relatively little action although its high altitude capability allowed it to be used effectively for reconnaissance.
9. See for example *The Last Year of the Luftwaffe* by Alfred Price, Frontline Books, 2015.

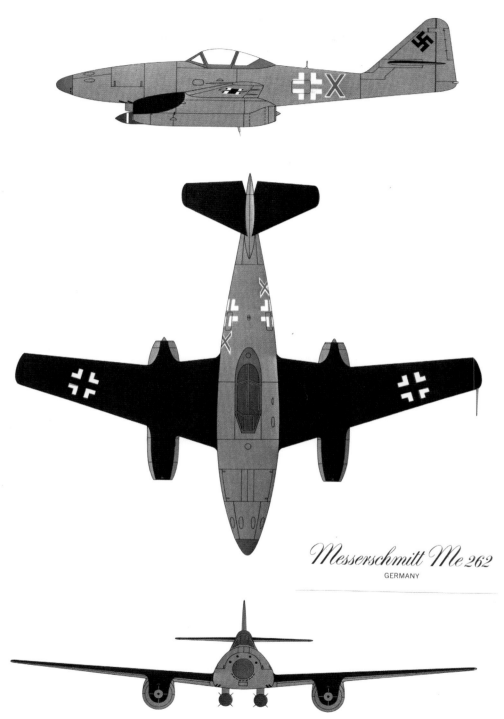

Fig. 34: Three-views of the Me-262 fighter aircraft (note the attachment of two bombs!) and the modest wing sweep.

New Aeroplanes – Allied Jets

During the war years, Whittle involved a number of industrial concerns in the building of his engines. The most effective of these relationships was with Rolls-Royce. It was at this time that the technology was shared with GE on the other side of the Atlantic. In 1944, as development efforts were bearing fruit, the British government took over Power Jets Ltd by nationalizing it. Whittle reluctantly agreed to this turn of events. In Britain and in the US, the application of the engine technology to an aeroplane was subsequently taken over by the military establishments.

Initially the goal was to build fighter aircraft. The jet engines proposed for these aircraft used radial-flow compressors that had very distinct disadvantages over the axial flow design employed in the Jumo 004. These included the difficulty of getting enough airflow into the engine and the shape of the engine; it was not what one might call 'slender'. The figure below shows the Rolls-Royce Nene, the third production engine by the company based on Whittle's work. The first and second were the similar Welland and Derwent engines.

The Welland engine powered the first British production jet aeroplane, the Gloster Meteor, version F.1. The Meteor was a twin-engine fighter aeroplane with rather 'fat' engine nacelles on the wings. It was put into Royal Air Force service in July 1944 and served in many other services in the years after the war. The F.1 version did see action in the war against the flying buzz bombs, the German V-1 cruise missiles attacking England, specifically London, during the summer of 1944.

The Meteor was ultimately developed to later versions, produced in large numbers (almost 4,000) until 1955, and successfully deployed all over the world. Figure 36 shows the three views of the F.4 version of the Meteor using Rolls Royce Derwent engines.

The first American production jet fighter (the Bell P-59) was also a twin, but dealt with the nacelle issue by mounting the engines at the wing root where some of the engine's bulkiness could be incorporated into the aeroplane's body. The engines in that aeroplane were GE I-16 (later called J31) engines (initially 1,650 lbs thrust, first run in 1943) that were based on the British design. The P-59 was not deemed useful from a military viewpoint and was not used in combat. Only sixty-six aeroplanes were produced.

The radial flow compressor engines in the first generation of British and American jet powered fighter aircraft made for interesting engine installations in the body or nacelles. Since the air into the compressor entered the two sides of a radial compressor (see Fig. 39, page 96), the space around the engine was a plenum of relatively low speed air to allow it to proceed from the aeroplane air inlets to the two engine air inlets on each engine. Typically this plenum was bounded at

Fig. 35: A Rolls-Royce RB.41 Nene engine (first run in 1944) museum display. The Nene used a double entry radial flow compressor. Half the air enters through the screen at right and a second screen just behind it and out of view admits the other half of the air used. The cutaway display shows details of the rear half of the compressor as well as the combustors' cans that are silver in the external view. The Soviet MiG-15 was powered by a copied version of this engine. (Photos: Hans Toorens, Historic Flight Foundation and author at New England Air Museum).

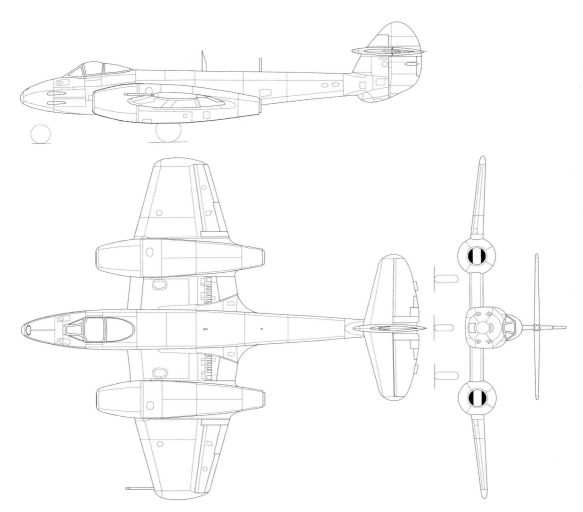

Fig. 36: Three views of the British Gloster Meteor F.4 jet fighter (Wikimedia Commons, Kaboldy).

the rear by a barrier, as shown on the tail pipe of the Rolls-Royce Nene (Fig. 35). Later examples of such installations can be seen in the Gloster Meteor, (Fig. 36), the Lockheed P-80 and the Grumman F9F, all members of the first post-war generation of fighter aircraft.

Such an installation contrast sharply with that of an axial flow compressor engine in that the motion of the air is maintained to keep the available pressure high. The intake flow is carefully tailored to smoothly meet the inlet guide vanes of the axial flow compressor engine. An example illustrating the direct entry of air into an axial flow compressor engine is included in Fig. 30 in a cross-section view, or Fig. 34 from an external viewpoint.

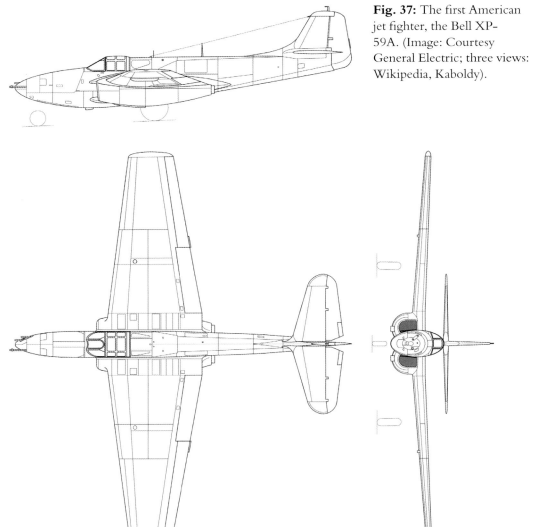

Fig. 37: The first American jet fighter, the Bell XP-59A. (Image: Courtesy General Electric; three views: Wikipedia, Kaboldy).

Fig. 38: The North American Aviation jet bomber B-45. (Peter M. Bowers Collection/Museum of Flight).

Jet Bombers

The jet engine did not play a pivotal role in the war but greater possibilities were certainly on the minds of the Allies. As the war was ending, American industrial aeroplane builders were contemplating building jet-engine-powered bombers. A competition to secure production of such an aeroplane was underway and contracts to build prototypes were awarded in May 1945.[10] North American Aviation built the XB-45 and Convair the XB-46. These aircraft utilized the jet engines of the

---

10. VE Day was on 8 May 1945.

day and used the current state of the art in design: a conventional, straight, high aspect ratio[11] wing. The X refers to experimental or prototype versions of the model and was dropped when the aeroplane was mass produced.

Boeing was also in that jet bomber competition with a proposed and yet-to-be-finalized B-47 design. All American engines in that period were designed by General Electric.[12]

General Electric and the J47

During the war years, GE developed engines with ideas from the British. Initial interest exploited the radial flow compressor, enlarged to produce an engine with about 4,000 lbs of thrust. This engine, the I-40 that later became the J33, found its way into the earliest of the American jet fighters, the XP-80 Shooting Star. Interest in new jet aeroplanes by the US military was great and production orders came to GE. Unfortunately, because of production capacity limitations, engine production had to be subcontracted to other firms, and the Allison Division of General Motors got into the business. While this radial flow compressor engine was being developed, a parallel effort to design and build an axial-flow compressor and an engine around it was underway. H.D. Kelsey at GE overcame the conservative forces in the organization that wanted to stay with radial-flow compressor technology. He got GE into the axial-flow engine business. The advantages of higher compressor pressure ratio and lower frontal area were just too good to pass up. A new TG-180 engine, that later became the J35, first ran in April 1944 and flew in 1946. Again, production capacity issues caused this engine to be produced by Allison.

These production contracts did indeed lead to Allison becoming a competitor and steps had to be taken to minimize the impact. GE proposed a new engine to the US Air Force: the TG-190. The proposal was accepted and the new J47 went into production in 1948 that ran until 1956. Thrust level was initially around 5,000 lbs and grew to almost 7,000 (without afterburning). It had a 12-stage axial flow compressor driven by a single-stage turbine. The compressor had a cylindrical external geometry with a pressure ratio of around five. Engine rotational speed was around 8,000 RPM. Eight relatively large can-type combustion chambers provided 1,600F gas output to the turbine. These parameters are cited here

---

11. Wing aspect ratio is a measure of its slenderness, defined as wing span squared ÷ wing area.
12. GE's involvement in the new gas turbine technology was based primarily on their liaison with the British development effort, which in turn relied on their strong experience with steam turbines superchargers, and turbochargers.

Fig. 39: A General Electric I-40 (4000 lbs thrust, an advanced version of the 1600 lbs thrust I-16) showing the double entry compressor as well as the various components in cut-open section. Note the turbine nozzle (partially cut away) and the turbine rotor. (Courtesy General Electric).

because they are good mile-markers to measure improvements made later in our story. The J47 was a major step forward in the evolution of the jet engine, albeit with only military applications, for now!

If a single image is to be chosen to illustrate just how difficult the development of axial-flow compressors was during the early years, the naked J47 shown below does it beautifully. The 12-stage compressor processes the same power as the single turbine stage. This implies that each stage of the compressor does relatively little, and achieving a compression ratio of five is a challenging matter. The steam turbine experience, and the fact that the pressure falls rather than rises as in the compressor, allowed the rather direct application of that technology to provide the turbine power output. Later, several decades of development of compressor aerodynamics also yielded good results inasmuch as each stage of a modern compressor raises the air pressure to a significantly higher degree.

Fig. 40: A GE J47. Note the slender geometry, the can combustors based on the Whittle technology, and the cylindrical compressor casing made in two halves. A single stage turbine is at the rear. Display model at the Museo Storico dei Motori e dei Meccanismi in Palermo, Italy. (Wikimedia Commons).

Seedlings

An important early builder of jet engines in the US was the Westinghouse Electric Corporation that, apparently without direct connection to British expertise, designed and built an 1,100 lb-thrust jet engine under contract for the US Navy during the Second World War. The company was well established in the steam turbomachinery business and thus had in-house expertise to tackle the challenges posed by the gas turbine engine. Testing of an engine with axial-flow compressor and turbine started in 1943 and progress towards higher thrust engines was made during the remainder of the war and thereafter. An improved, higher thrust engine, to be called the J30, went into production for the first navy jet fighter aircraft, the McDonnell FH-1, Phantom.

Westinghouse and the US Navy granted a licence to Pratt & Whitney to produce the J30 engines. Westinghouse's time in the jet engine business was relatively short. It went on to produce a number of more advanced engines for the navy, but closed the Aviation Gas Turbine Division in 1960. The company can claim credit for building and running the first American axial compressor jet engine, the J30, and its follow-on, the J34, used in military aircraft. Westinghouse jet engines never powered any production airliners, but it likely helped sow the seeds for the business at Pratt & Whitney.

The Phantom was introduced into navy service in 1947 and retired in 1949. It is likely that rapid technological advances rendered this aeroplane and its engines obsolete in a rather short time. However, industrial interest in gas turbine engine technology was indeed widening.

Chapter 5

Cold War Engines and Post (Hot) War Airliners

Technology Transfer

German technology has a place in American jet engine history. The same can be said of that history in France and in the Soviet Union. During the war, the author's father worked at Junkers in their jet engine group designing the control system for the Jumo 004. The Junkers company was located in Dessau where the region was to become the Soviet Occupation Zone. It was our family's good fortune (as well as that of the other engineers employed at Junkers) that Dessau lies on the west side of the Elbe River, the line that US troops used as the stopping line[1] of their march east. The American contact exposed the engineers at Junkers to American troops and, eventually, yielded invitations to retreat west with them as they ceded the area to the Russians as their occupation zone.

There were already Cold War tensions between the US and the Soviet Union and both parties tried to maximize their harvest of the intellectual loot represented by the builders of the first production axial-flow turbojet engine and aviation technology more generally. The Junkers engineers were interviewed by Soviet, French, and American officers and were offered a chance to follow any of them to where they might wish to work. These meetings were cordial, according to those who participated. Many of the engineers chose to go to the West, primarily to the United States. A number who did not accept American or French invitations to be evacuated to the West became the kernel of the new gas turbine industry in the Soviet Union. This group was initially kept at Junkers in (East) Germany but was ultimately forcibly taken to Russia.

The line drawn by the Allies as the demarcation between the Soviet and Western occupation zones held a number of industrial concerns that were centred on the development of turbojet engines. BMW had people and facilities in Stassfurt, near Magdeburg. Technical staff in Dessau who did not accept offers to go west as the

---

1. US General Eisenhower stopped his advance eastward to avoid accidental conflict with the Russians moving westward and he saw no point in losing American lives to conquer territory that would ultimately be occupied by the Soviet Union, as per the Yalta Conference agreement.

US Army ceded the territory to the Soviet Union for occupation on 1 July 1945 were reassembled by the Russians. Their task was to develop the BMW 003 and the Jumo 004 to higher performance without access to much in the way of the materials needed, specifically nickel and chromium, as well as access to technical information and facilities. It was a rough start. The engineers, nervous about Soviet intentions, were reassured that they were working on a joint Russian-German enterprise that was to remain in Germany.

On the night of 21 October 1946 however, these promises were broken, with the transport of about 2,000 people and everything associated with jet engine development to Russia. 'Everything' included all personal possessions and even a female friend of one person whose wife was not present at the time of the roundup!

The work of these engineers ultimately led to the construction of jet and turboprop engines used by the Soviet Union in military and civilian aircraft. The large counter-rotating turboprop engines powering Soviet aircraft (the Tu-95 bomber, NATO codenamed Bear, and the civilian transport Tu-114) were the product of these engineers, especially so after the 'firm' associated with the designer Nikolai Kuznetsov took over to produce a number of engines with NK numbers, the famous NK-12 among others. The German engineers were ultimately allowed to return to Germany between 1950 and 1953. A footnote is that they later tried, unsuccessfully however, to build a jet airliner, the Baade 152. That sad story, with its political subtext, is a monument to the fallacy of a long and successful life in a planned economy with controls exercised from Moscow.

The jet engine group from BMW was offered a chance to work in France. The French intent was to establish a jet engine industry where none existed. Due to the fortunes of war and its aftermath,[2] the author's father accepted a position there.

The American Operation Paperclip sought to secure as much information of various kinds that might be useful in the still ongoing war in the Pacific. This operation had two dimensions relevant to this story, one was the visit of American experts to factories and research institutions in Germany to review technical work that was in progress as the war ended. The other was the removal (by invitation) of the principal people involved in jet engine technology from Germany. Both of these aspects had consequences for the aviation industry in the US and elsewhere. They also shaped the life of the author. The approximately 1,600 Germans and Austrians on a roster of Operation Paperclip, dated January 1947, included

---

2. A serious truck accident during the evacuation by the US Army. The resulting chaos caused the papers to join others headed to the US to have been lost. American urgency was probably also diminished by victory over Japan by the time the accident issues were resolved.

Fig. 41: The Tu-95 'Bear' bomber of the Soviet air force (wikipedia).

scientists and engineers with expertise in many disciplines. The author's father's name was on that list. Another name on the list was that of Adolf Busemann who originated the wing sweep idea. In 1947 he emigrated to the US and worked for the National Advisory Committee for Aeronautics (NACA) that later became the National Aeronautics and Space Administration (NASA), and taught at the University of Colorado.

First, the Allied scientists in Germany. One particular American team with a special interest in matters aeronautical visited a German facility in Völkenrode.[3] The team was led by Theodore von Karman and included The Boeing Company's chief aerodynamicist George Schairer. His journey yielded the finding that wing-sweep was being investigated on wind tunnel models and on prototypes of new aircraft. In his book *The Road to the 707*, William H. Cook notes that the information from Schairer came to Boeing in May 1945. It changed the design of the B-47 from straight to swept wings. Because Schairer's role involved service to the US government, the wing-sweep data was shared with other manufacturers but ignored. That was the crucial moment that again put Boeing on the map with an innovative aeroplane design to shape almost all future aircraft, military as well

---

3. The site of the Aeronautical Research Institute, LFA, also known as the Hermann Göring Research Institute.

as civilian. The B-47 was significantly faster than the other XB bomber prototypes being evaluated by the US Army Air Forces, shortly to become the US Air Force.

The people evacuated to the US included a number of the team that built the Jumo 004, specifically Anselm Franz, and Hans von Ohain. They were initially put to work at Wright Field in Dayton, Ohio, where von Ohain finished his career as a staff scientist at the US Air Force and later taught at the University of Dayton. The Junkers engineers under Franz wanted to build engines. In 1952 they obtained a US Army contract to build a facility whose initial purpose was to supply turboshaft engines for the army, principally for helicopters. Thus AVCO-Lycoming's Gas Turbine Division became part of the engine manufacturing community in the United States and the door opened for the author's father and his family to rejoin former Junkers colleagues there.

New Business for Pratt & Whitney

The post-war period would determine who, among companies, was going to be in the jet engine development business. While the entry route into the jet engine business by GE and Rolls-Royce is relatively clear, Pratt & Whitney's history is more complicated. They had not participated in development work on these new engines because the US government thought that their continued focus on producing reciprocating engines was necessary for the war effort. After the war, the company made the financial commitment to participate in the switch to jets. Pratt & Whitney gave itself five years to get into the business and do it with engines that leapfrogged the competition.

The company navigated the post-war period in interesting ways. For example, GE and Westinghouse were eager to secure production capacity for their radial-flow (GE J33) and axial-flow engines (Westinghouse J30) that these companies had sold to the US military. They approached Pratt & Whitney for their production capability as licensee for the engines. Frederick Rentschler and Leonard Hobbs (corporate vice president) turned them down because they wanted Pratt & Whitney participation in engineering development to be part of the deal and a pure manufacturing contract had less value for developing Pratt & Whitney's future business. In the negotiations it was important to consider the needs and views of the ultimate customers, aeroplane manufacturers and the US military. The US Navy's need for the J30 engines ultimately led Pratt & Whitney to accept an agreement to produce these engines. The benefit was customer relations with the Navy and the opportunity to learn about axial-flow compressors that Pratt & Whitney believed were the better compressors for future engines.

The aeroplane builder Grumman was a good customer for Pratt & Whitney engines and they needed engines for a fighter jet that ultimately became the F9F, Panther. Grumman wanted the Rolls-Royce Nene engine for this aeroplane. Pratt & Whitney was approached by Rolls-Royce to manufacture and 'Americanize' it. Naturally, Rolls-Royce interests were rooted in entering the US market. Pratt & Whitney should have categorically turned them down. However, Navy and Grumman considerations as well as a Rolls-Royce commitment to involve Pratt & Whitney in the development of advanced versions of the Nene allowed the licensed production of the Nene as the J42 to proceed. The follow-on development engine became the J48. The experience gained by the production of these engines (J30, J42 and later J48) helped establish Pratt & Whitney as a competent producer of jet engines.

The Korean War (1950–53) was a time to test the capabilities of jet engines in military combat. An important aeroplane in the US arsenal at that time was the Lockheed F-80 powered by the radial flow Allison J33 engine. The time was, however, to be the end of engines with radial-flow compressors used by the US military for first line combat aircraft. The US Air Force was already using the Republic F-84 with the Allison J35 and Wright J65 engines and the North American Aviation F-86 with GE J47 engines. These last three engines had axial-flow compressors.

It was also a time when wing sweep in an aeroplane configuration was recognized for its speed advantage of 50–100 mph. For example, the F-84F was a swept-wing version of that aeroplane whose many other versions had straight wings. Similarly, the US Navy's J48-powered F9F-9 Cougar was a swept-wing variant of the Panther.

The adversaries to the UN forces in the Korean conflict were the North Koreans backed by China. They used Russian Mikoyan-Gurevich designed fighter aircraft, the MiG-15 and MiG-17. Both used Klimov engines developed from the Rolls-Royce Nene; the technology was perhaps stolen from or, reluctantly, shared by the Western powers with their Soviet ally during or after the Second World War. It is a bit ironic that the technology used in the Korean air by both sides in that conflict originated from the same source. Technology is held within boundaries with great difficulty.

Hopping over the Ocean – not

At the Second World War's end, all bombers and transport aircraft were powered by piston engines and propellers. These engines had limited power and the aircraft had limited range. They could barely, if at all, cross the Atlantic from North America to Europe. Refuelling stops were the order of the day. Willy Ley, the German-American space and aviation enthusiast and author, suggested that large artificial floating islands could be placed in the middle of the Atlantic to refuel aeroplanes and refresh passengers and aircrews. Fortunately, it didn't have to come to that. Exposing aircraft to the weather at sea level would have been problematic and the time for the journey from the US to Europe would have been quite long, at least when seen from the modern perspective of flying non-stop at 550 mph. Today, flying from New York or Los Angeles to Europe, airliners use great circle routes into the arctic regions where the positioning of Ley's refuelling islands would have been difficult if not impossible. The alternative of staying in the milder latitudes would have involved more travel miles.

The Last of the Piston Engine Airliners

The war years required new and capable transports. Douglas, starting in 1942 built the DC-4 with the military designation of C-54, 'Skymaster'. It was produced in large numbers with 1,150 HP engines. Douglas continued military production of transports with a new C-118, 'Liftmaster'. It was a larger and pressurized transport. A civilian version, called the DC-6, was produced after the war as higher power engines became available. It was faster with its 2,400 HP engines (300 mph versus 220 mph) with roughly the same seating capacity as the DC-4. These four-engine aircraft played a big part in the Berlin Airlift in 1948–49.

Douglas commercial aeroplane history includes the DC-5, a twin-engine, high-wing short-haul airliner. It unfortunately was caught in the events surrounding the start of the Second World War and never was produced in large numbers either as a commercial or military transport. William E. Boeing owned a DC-5 for his personal use. His choice for an aeroplane was not a Boeing-built airliner. That may have been a reflection of William Boeing's feelings about the company he had created earlier and left.

During the war, Lockheed was developing a model L-049 for Transcontinental and Western Air (TWA, renamed Trans World Airlines in 1950) and for Pan American Airways when the War Department transformed the Lockheed effort into a transport to become the C-69. Difficulties with the aeroplane and its engines resulted in it being produced in small numbers as a military transport. It was, however, to become a major airliner after the war when it was recast as such: the 'Constellation'. This aeroplane was designed by Clarence 'Kelly' Johnson in what later became the 'Skunk Works', the birth-place of many ingenious military aircraft under his leadership. The Constellation was a beautiful aeroplane,

Fig. 42: Lockheed L-1049G Constellation over New York City. (Museum of Flight Collection). Lower: Wing tip tanks for extended range on a TransCanada Air Lines 1049 Super G. (Photo by author at the Museum of Flight).

with its nose bent downward to shorten the landing gear and the rear swept up to avoid ground contact on take-off. There was little size and shape similarity between the fuselage body frames from front to rear. This aspect of the design led to issues when more powerful engines allowed for a heavier aeroplane and fuselage lengthenings were not readily accommodated without some compromise on 'beauty' or production cost. One fuselage lengthening was indeed made on the aeroplane but the cylindrical shape of competitive aircraft made such alterations much easier.

Airlines also began thinking of new aeroplanes. TWA for example investigated the feasibility of aeroplanes significantly larger that the DC-6 and the like. Under consideration were a version of the Hughes H-4 (called the HK-1, the civilian name for the *Spruce Goose*), the Lockheed Model 37 'Constitution', a civilian version of the Convair XC-99 which was itself a transport version of the B-36 bomber, and a new DC-7 that was a version of the C-74 'Globemaster'. The Lockheed and Convair aeroplanes were to be double-deck aircraft with a capacity close to 200 passengers. The problem was the engines. These aeroplanes were going to use the Pratt & Whitney R-4360, but even these most powerful engines were short on power and had limited growth potential. In the end, a realistic look from a marketing viewpoint concluded that the industry was not ready for large aeroplanes. None of these aeroplanes went beyond prototype construction, if that.

The airframe industry took their experience with military transports and built transports with similar performance for civilian service. The Douglas DC-6, the Lockheed L-049 and later L-649, and the Boeing's Model 377 'Stratocruiser' (introduced in 1949) rounded out civilian airliners built on the west side of the Atlantic in the post-war years. Boeing based the design of the Model 377 on the B-29 and the C-97. The 377 used the Pratt & Whitney R-4360 rated at 3,500 HP. The fuselage was of a double tube design[4] (see Figs. 23 and 44) with a lounge at the lower level and seating or berths in the upper. The DC-6 ran with Double Wasp engines and the L-649 used the Wright R-3350. The power rating from the latter two power plants was about 2,500 HP.

The war taught one difficult lesson resulting from the use of piston engines. There was not much design freedom in choosing where to mount the engines and their propellers. The engine had to be high enough so that the propeller cleared the runway surface safely, and for a multi-engine aeroplane, the location had to be on the wings. Just about all engine mountings for bomber and transport aircraft were at the leading edge of the wings (tractor propellers). One notable exception was the Convair B-36 where the engines were at the trailing edges (pusher propellers). In a combat situation, damage to an engine was often very serious because the high-octane gasoline the piston engines required could catch fire. If not dealt with successfully, this could burn the wing resulting in loss of

---

4. Lockheed designed and built two military R6V Constitution prototypes that also used the double tube fuselage. They flew in 1946.

the aircraft. In civil aviation, such events are much less likely than in military combat but the probabilities are not negligible and it was a risk neither airline nor aeroplane manufacturer enjoyed having to live with.

Thinking along these lines led The Boeing Company to design the B-47, their Model 450, with the engines on struts hanging below the wing. In that location, an engine fire could be more easily contained and was less likely to involve the wing. In a dramatic case of major mechanical failure, the engine could 'simply' fall to the ground and the aeroplane proceed safely. An additional dimension of the jet age is that so-called jet fuels are much less volatile than gasoline and that makes using them inherently safer.

Airliner Competition

In the 1950s, the competition for airliners demanded by airlines intensified as the post-war economy improved. Douglas and Lockheed, for example, were head to head in competition for the airline business. Opposing offerings included: the DC-4 vs. L-049, DC-6 vs. L-649, DC-6B vs. L-749, 749A, and 1049, DC-7 vs. 1049G, and DC-7C vs. 1649A. Each phase of the competition took advantage of the newest engines available. Woven into the mix were the 307 and 377 models

Fig. 43: An American Airlines Douglas DC-7. (Peter M. Bowers Collection/Museum of Flight).

from Boeing. The goal was speed and reach, first transcontinental and then transatlantic.

The range of the Lockheed Super G 'Connie' (L-1049G) had, while the design was called the L-1049E, been short of the capability of the DC-7. Improvements were made, including the addition of external fuel tanks at the wing tips. The extra fuel tanks solved the range problem and made the 1049G the fine airliner that it became. The acquisition of these aeroplanes by TWA and the role of Howard Hughes, the majority shareholder of TWA, in that process turned out relatively well. It became, however, an overture to much greater turbulence when the piston engine aeroplanes were to be replaced by the jets.

The evolution of the Lockheed Constellation family of aeroplanes was in no small part determined by TWA. The TWA management and Howard Hughes were often at odds over procurement issues and the aeroplane makers. In his book *Howard Hughes and TWA*, Robert Rummel, who enjoyed a close and trusted relationship with Hughes, describes the turbulence associated with TWA's purchase of a number of aircraft, including the Super Gs and short-range twin-engine airliners. Rummel was the chief engineer at TWA and had a major say in the procurement of airliners for the company and determining how they were used.

Lockheed built and tried to sell ever more capable Constellations. One of the configurations was the L-1249 prototype ordered by the US Navy equipped with 5,500 HP Pratt & Whitney T34 turboprop engines. Four of these were built in 1954. A proposed L-1449 airliner consisted of the 1049 body powered by the T34 turboprop engines. This design was an adaptation of the L-1249 for civilian use but no commercial sales were made. In a time of growing need for greater passenger capacity, the aeroplane was too small for TWA and the engines were not ready. Over TWA objections, Hughes nevertheless signed a vaguely worded contract for purchasing the L-1449. The Pratt & Whitney engines turned out not to live up to expectations in terms of performance and projected reliability. In part, they were thought to have required an overly heavy protection system in case of compressor blade failure that was, at that time, a perceived risk. Ultimately Pratt & Whitney withdrew the engine and the L-1449 programme went into limbo in January 1955, to TWA's relief. The gas turbine engines were, however, nibbling at the opportunity to be on airliners. The L-1449 was salvaged as a project and renamed L-1649G (Starliner) with an improved version of the Wright R-3350 turbo-compound engines. It had a new wing and was an effective competitor to the DC-7C. Forty-four of the L-1649G were produced between 1956 and 1958.

The Lockheed aeroplanes and the Boeing Stratocruiser cruised at speeds near 300 mph while the DC-7 was claimed to cruise at 359 mph. For a time

Fig. 44: Boeing model 377 Stratocruiser in flight. (Boeing Images BI23736)

in 1951, Howard Hughes and TWA considered ordering their Lockheed 1049 Constellations with additional Allison J35 jet engines to boost speed. They wisely dropped the idea when the fuel consumption impact on range became clear. It is also safe to say that jet engine technology was still a risky proposition in the civilian aviation sector.

Airliners in this time had a capacity of around 100 passengers and ranges that approached being called intercontinental. The DC-7C model, for example, had a range of 5,600 miles, well able to span the 3,600-mile trip from New York to Paris. The DC-7 and the L-1049G Super Constellations were built in numbers of roughly 350 (121 DC-7Cs) and 250 respectively, when production ceased in 1958. Both used the Wright R-3350 'Duplex Cyclone' turbo-compound engines with a power output of about 3,300 HP. Fifty-five Boeing Stratocruisers were built.

The years before the last of the piston engine airliners were built were heavy with the anticipation of airliners powered by jet engines. There were many important and open questions that were resolved with the passage of time. Front and centre was the question of range. Fuel consumption by jet engines was a significant concern. The speed issue centred on the question of whether turboprops were going to be leapfrogged to reach the penultimate state of near-sonic cruise speed in jets. That question was important because it determined the planning horizon and the need to smoothly integrate the new aeroplanes into existing fleet operations.

The engine industry was delivering good jet engines to the military. Were these engines suitable or even available for commercial use? At Boeing a question was: would the US Air Force permit the design and tooling for the KC-135 tanker to be used for simultaneous production of the 707? Lastly, how much time was required before the jets would find their way into service? These questions quickly evaporated and jet aeroplanes became the new reality.

A vexing dimension of the TWA transition to jets story centres on Hughes's unorthodox management style (if 'managing' is the right word) of what he owned or controlled, the Hughes Tool Company and TWA. That style hampered TWA from moving forward smoothly. While it is difficult to characterize Hughes' role, it can be said he always sought a competitive advantage for TWA over rival airlines. To that end, he tried to involve a number of airframe builders in schemes where TWA would be the first and sole provider of the service for a long time after its introduction. Companies that were involved in design exercises with TWA include the British firms A.V. Roe and Vickers, Avro of Canada, Convair (including the attempt to transform the B-36 into a commercial airliner) and, of course, Lockheed. The Hughes approach was to monopolize the initial production run to exclude the competition. It can be said that these ventures did not end well, at least insofar as the airframe companies getting into a long life of producing airliners was concerned.

As these exercises were going on and shortly after the rollout of the Boeing Dash 80, salesmen from Seattle were promising 707s for delivery in 1958!

Howard Hughes' need for monopolizing initial production runs notwithstanding, Boeing and Douglas fully understood the need for a sales and delivery regimen that maximized their access to as many airline customers as possible and to satisfy the needs of these customers. They certainly did not need a delivery process that sent airlines over to the competition because production commitments to one customer extended delivery for others into the distant future! And so it was that Hughes failed in his bid to secure a unique and advantageous fleet for his TWA by building aeroplanes.

The last big Piston Engines

The last of the big airliners and their engines illustrate the two ways that two different manufacturers tackled the problem of obtaining 3,500 HP. Pratt & Whitney made the displacement of their R-4360 Wasp Major large. On the other hand, the Wright Duplex Cyclone R-3350 was configured to make the (smaller) basic engine harness the power in the exhaust gas. They built what is known as a turbo-compound engine. By this method, three exhaust turbines transfer power to the engine crankshaft through fluid couplings while the supercharger is mechanically gear-driven. Both methods were effective. The Duplex Cyclone weight is given as 1.2 lbs/HP, somewhat heavier than the Wasp Major (see Table 1, page 73) but possibly better in terms of fuel consumption. This engine was the first American aircraft engine built with direct fuel injection technology. Pratt & Whitney had, in the early 1930s, tried fuel injection on their Wasp engines but reverted to use of well-developed carburettors. C.F. Taylor suggests that Germany successfully used fuel injection on their military aircraft in the Second World War.

It is interesting to note that a turbo-compound engine can be thought of as a turboshaft engine with the internal combustion engine as the heat source. The ICE naturally produces a large amount of mechanical power but also a lot of thermal power delivered at pressure sufficiently high to be useful. Seen in this light, the turbo-compound resembles an ordinary gas turbine.

Short-Range Twins

In addition to the large airliners built and operating after the war, there were manufacturers who built smaller airliners for the regional market. The Glenn L. Martin Company was an active producer of military aircraft but also built a small (ca. 40 passengers) twin-engine airliner that used the engines of the day: Pratt & Whitney R-2800, Double Wasps. An unpressurized Model 2-0-2 (with tricycle landing gear) was designed and built shortly after the war to compete with the DC-3. A serious structural design issue was revealed by a fatal accident. In August 1948, a 2-0-2 suffered a wing failure that was subsequently detected in other aircraft as imminent. After the fleet was grounded and the necessary repairs made, the aeroplane was designated as Model 2-0-2A. A Model 4-0-4 with cabin pressurization was built to replace it in the early 1950s. About 100 of the 4-0-4s were built. Both models cruised at just under 300 mph and had relatively short range. They were Martin's last venture into the airliner business.

A more successful American builder of the post-war DC-3 replacement aircraft was Convair who introduced the CV-240 in 1947 and produced well over a thousand aircraft for many client airlines until 1954. The pressurized cabin CV-240 also used Pratt & Whitney R-2800 Double Wasp engines to carry forty passengers on short range flights at over 300 mph. It also had tricycle landing gear and was developed further, albeit in smaller numbers; improvements included turboprop engines and greater passenger capacity.

Among the British twin-engine aircraft that were sold widely is the Vickers VC.1, Viking. It was quite similar to the DC-3 in its outward appearance and capability: thirty-six passengers cruising at 210 mph.

Among the unsuccessful competitors in the short-range commercial twin market after the war were Boeing with a proposed (and never built) Model 417 and Curtiss-Wright with a Model CW-20 that was built in very large numbers as the C-46 'Commando' for the US military.

Time between Overhauls

Airlines always show smoothly running, beautiful aeroplanes in their sales media to attract and comfort customers. The promised smooth operation was not always realized in practice however. Much labour and money was required to make the reciprocating engines do what they were built for and they did not always perform. Viewed from well into the jet age, one can say that the very nature of the internal combustion engine is partly to blame for its own demise.

Consider that radial aircraft engines had to have inspection visits about every fifty hours. After four such inspections, that involved various visual checks, filter changes, and refits in the shop, the expectation was that the engine was about half way to a complete overhaul. This would imply that an engine would be taken apart and reassembled with new parts fitted where necessary after about 400–500 hours of service. A Pratt & Whitney manual for the R-1830 in 1943 military service stipulated 350 hours, which could be stretched to 600–1,000 hours if duty was light and involved long-range operations. The key determinant to establish need for overhaul was oil consumption, which had to be carefully monitored.

These engines were fairly heavy oil consumers and the aircraft had sizable oil tanks near the engines to satisfy the need. For example, in the Second World War-era Pratt & Whitney R-1830, about three percent of the weight of consumables was in the form of oil. Even when new, these engines were built with oil seal rings on the pistons that allowed oil to generously bathe the cylinder walls. Anyone who has ever seen a large radial engine being started will notice two things related to oil. The first is that the propeller is turned before ignition is initiated. This is to insure that oil leakage into the lower cylinders will not cause 'hydraulic lock' resulting from the piston attempting to compress the oil that might have pooled there. Turning without ignition ahead of starting was to allow the oil to leave the cylinder slowly. If the engine did indeed lock up, the oil had to be drained by other means. Failure to allow for this would likely have resulted in the dramatic removal of the cylinder head! The second item to notice is the heavy oil-combustion smoke-emission at engine start. This is the result of the oil caught in the cylinders at the start of operation. The role played by lubricating oil in our modern automobiles is far tighter, for air pollution and operational reasons, at the very least.

Mellberg (see bibliography) suggests that the lack of success of the Boeing 377 as an airliner can be attributed to the 'temperamental and costly engines'. A little more descriptive are the words of the late Pan Am chief flight engineer John S. Anderson who stated, 'the mean time between overhauls was 500 hours – *if you were lucky*! In the summertime, Pan Am averaged two engine failures per day and had to hold twenty percent of its 377 fleet in reserve, to pick up passengers, and to ferry new engines, engine parts, mechanics, and tools to 377s that had lost an engine.'[5] It is not hard to speculate that the problem with the Pratt & Whitney R-4360 engines was the marginal cooling of the last one or two rows of cylinders. They were cooled with the warm air from the cylinders ahead of them. This issue was surely exacerbated during the summer season.

Other airliners powered by the last and most powerful engines, including the Wright Duplex Cyclone, did not escape the travails of those described above for the R-4360. The engine manufacturers pushed the limits of what was possible very hard, always hoping that TBO (the engineer's acronym for 'time-between-overhauls'!) could eventually reach 2,500 hours. Indeed, by appropriately rating the power output of small civil aviation engines like the Continental O-470 described earlier, the TBO for such engines approached 2,000 hours.

The jet age experience with TBO was exemplified by K.F. Whatley who characterized the improving experience with P&W JT3C and JT3D engines on Boeing 707s and 720s in the period shortly after introduction of service in 1959 with American Airlines. His paper, presented at an American Society of Mechanical Engineers conference in March 1962 (see bibliography), indicates that their experience took TBO from 800 to over 2,100 hours in two short years as the company learned to operate and manage the new jet engines in their fleet. The paper outlines the nature of the teething problems associated with the new engine type, all resolved as time passed.

---

5. Personal communication, John L. Little, Museum of Flight, Seattle.

To put these numbers of hours in a later context, we note that in 1985 the JT8D turbofan was subject to engine inspection every 5,000 cycles of take-off/landing. That amounts to 10,000 hours of flight service if one assumes that the average flight length is two hours. A rebuild visit to the shop was not expected to be before 10,000 cycles.

Today, the time between overhaul exceeds 20,000 or even 30,000 hours and may be called for only when on-board engine diagnostics reveal a reason to look at a problem. Since the experience with engines and their durability is so good, it begins to approach the situation that an engine may *never* have to be overhauled! Just scrapped.

The Turboprops

The gas turbine had, in the time when jet powered airliners were under serious scrutiny, made inroads into commercial aviation, in the form of the turboprop. This engine consisted of a gas turbine with all the power produced sent to a gearbox that turned a propeller. Its virtues included greater power than the piston engines, lower weight, and lower operating costs.

The technology was most advanced in Britain where the Vickers Viscount was introduced into service in 1953 and produced until 1963. The Viscount was a 4-engine medium range aeroplane with 2+2 seating that started out carrying 32 passengers. With engine upgrades and body lengthenings, passenger capacity increased to 75. The engines were the Rolls-Royce Dart turboprops in versions that ranged from 800 to almost 2000 HP. The approximately 450 Viscounts made were widely used by airlines around the world. In spite of being a four-engine aeroplane, it competed well with the twin piston-engine aircraft of the day. It owes its success in part to the nature of the gas turbine engines: quiet and smooth with a promise of greatly reduced maintenance costs.

The Rolls-Royce Dart (like the Nene, named after a river in England), also adorned with the less romantic name of RB.53, was used in a variety of smaller airliners such as the Fokker F-27 and the Hawker-Siddeley HS-748 and produced in large numbers. Production of the engine ended in the late 1980s. It is noteworthy that this engine was produced until this recently because it had a two-stage radial compressor that one might have relegated to history as old technology because of the tortuous airflow through it. The radially outward output of the first stage had to be slowed and redirected towards the engine centre-line to be run through the second compressor stage. Awkward but effective. The reality was that it worked well.

The power limitation of piston engines and the maintenance cost issue led only Lockheed to offer and sell an American turboprop powered airliner, the L-188, also named 'Electra', as was the Lockheed Model 10. It was introduced into service in 1958 with four 3,750 HP Allison 501 turboprop engines. The engines were located above the straight wings and had rather wide-chord propeller blades to be able to take advantage of the power available. The aeroplane was a good performer. It cruised at 373 mph with 98 passengers in 2+3 seating. Unfortunately it suffered from a vibration resonance in the outboard engine mounts that resulted in several fatal losses of aircraft, two in 1959. The design flaws were successfully fixed but at a high cost. Market uncertainty and the advent of jet airliners limited production (ending in 1961) to 170 aeroplanes, some of which were of the military model, the US Navy's patrol aircraft, the P-3 Orion. The limited success in producing a Lockheed turboprop airliner did not prevent the company creating a very successful, somewhat similar, military turboprop high wing aeroplane, the C-130.

The turboprop airliner did find a footing in the Soviet Union, but only for a time. The Tu-95 bomber was a reasonable basis for modification as an airliner. In

Fig. 45: A Varig (Brazilian Airline) Lockheed L-188 Electra. Note the turboprop engine installation forward and above the wing and the wide chord propeller blades. (Photo: Helio Salmon taken in Buenos Aires).

the mid-1950s a large modern airliner to rival western aircraft was needed. The logical approach was to take the wings, the counter-rotating propeller propulsion system, and the main landing gear and mount them on a new body. In contrast to the mid-body wing mounting on the bomber, a low wing installation on the body allowed for an unobstructed cabin. The design was the Tu-114. It used the same almost 15,000 HP NK-12 engines as the bomber. The swept wing allowed it to fly fast and cruise at Mach numbers around 0.7. This was slower than the jetliners built by Boeing and Douglas, but quite respectable. The aeroplane was also the largest of that time, with a passenger capacity of up to 220. Finally, the generous fuel capacity gave the aeroplane a range that was also notably long, over 6,000 miles. This aeroplane was so capable that it established a number of world performance records that stand to this day.

Thirty-two Tu-114s were produced in the years 1958 to 1963 and operated by Aeroflot, the Soviet Union's national airline. An unusual arrangement between Aeroflot and Japan Airlines allowed the aeroplane to be used in service between the respective countries between 1967 and 1969.

In the end, the propulsion system did influence the destiny of the aeroplane. It was noisy, both inside the cabin and in the airport environment. This issue became an important dimension of its future when the jetliners were more widely used and thought to be better from noise and other perspectives. The type was retired in 1971.

Speed, again

Whatever the maintenance cost and complexity issues were with piston-engines, there was one important parameter that sealed the fate of not only the piston-engine in commercial aviation but also that of the turboprops: the propeller itself limits the speed of the aeroplane. Commercial airliners in the late 1950s travelled near 350 mph and not much faster without incurring fuel use penalties. A military propeller-driven fighter aircraft might be able to add another 100 mph to the top speed, but that was still short of the speed of sound near 700 mph.

The tip of a propeller cannot be allowed to exceed the speed of sound. The tip speed is the vector sum of propeller rotational speed and forward flight speed. The physics of the speed limitation is related to the same phenomena that allowed superior speed performance from swept wings compared to that of straight wings. It is clearly impossible to have flight speed approaching the speed of sound (as our present-day jet airliners do) while the tip speed is less than sonic. Aeroplane designers had to conclude that higher speed airliners would require the use of the jet engine. Increasing the engine power to the propeller would not solve the challenge of economical flight at higher speed.

In time, however, the propeller would be reinvented as a 'fan' with a duct around it to slow the fan's 'flight speed' down and thus allow a new kind of propeller-driven aeroplane to operate near the speed of sound.

Rotary Wing Aircraft

The engines that powered the big aeroplanes proved useful for helicopters. These aircraft, capable of hovering, were thought to have military as well as civilian potential by pioneers like Igor I. Sikorsky. As interest in flying boats waned, he turned to the development of this type of aircraft.[6] He called them 'Direct Lift Aircraft' in the successful patent applications he made starting in 1931. These patents outlined the basic ideas for controlling the thrust and the direction of the lifting system and established the basic configuration of the single lift-rotor with a tail rotor for rotation stability. During the Second World War (1943), he secured contracts to manufacture such helicopters for the US military. The R-4 and R-6 aircraft were of modest capability and were used primarily for observation and light transport. They did not play a pivotal role in combat operations as helicopters would in future conflicts. After the war, he and his Sikorsky Aircraft Company went on to design and build a number of ever more capable helicopters that continued the S- numbering system that was descriptive of the Sikorsky flying boats.

Powerful air-cooled radial engines made larger helicopters practical. Because the field was full of promise, Sikorsky was joined by other manufacturers, including Bell, Kaman, Hiller, and Piasecki. Of these, Piasecki ended up being the only other builder of large helicopters that had the potential for being used as airliners. The company was known as Vertol for a time and was ultimately purchased by Boeing. Frank Piasecki's (1919–2008) work was the basis for the tandem rotor design of aircraft like the Chinook.[7]

Sikorsky's post-war construction successes included the S-55 (1950) and S-58 models (1954) that featured the last significant use of a radial piston-engine in a large helicopter. The jet age had dawned.

---

6. The timeline of his work on helicopters is illustrated by a patent application entitled 'Direct Lift Aircraft' on 27 June 1931 that was granted as US1994488A in March 1935. A follow-on US231825A submitted under the same title was granted in 1943.
7. F.N. Piasecki, 'Tandem Rotor Helicopter', US 2,597,993 filed 16 Dec. 1946, granted 16 May 1950.

Fig. 46: Top: the engine installation on a Sikorsky S-55 helicopter. Bottom: This 12-passenger S-58 airliner served in the airports of New York, Los Angeles, and Chicago, as well as with Sabena Airlines as the livery reveals. The engine installation is similar to that of the S-55. (Igor I. Sikorsky Historical Archives ©2018).

B-47 – A large, fast Jet-Powered Aeroplane

American competition to build a long-range jet-powered bomber is an interesting story because it illustrates the transition that was required in thinking about a number of important aspects of aeroplane design. Boeing was a participant in the bomber competition and was up against proposals from North American Aviation (a B-45 based to some extent on the B-25, see Fig. 38), Convair (XB-46), and Glenn L. Martin Co (XB-48).[8] These last three aircraft were straight winged and powered by General Electric/Allison J35 axial flow compressor engines. All engine pods were partway out on the wing and closely integrated with it. Of these aircraft, the Convair XB-46 had nacelles that were quite voluminous because they also housed the main landing gear. The North American proposal won an initial round and was awarded a modest production contract for aeroplanes with J47 engines. Boeing participated in a second round competing with Martin and had to come up with a winning design. It is not the intent to tell the entire B-47 story here because that is well told in Cook's description of events. In short, Boeing won the competition with the swept-wing B-47 because it paid attention to the swept wing research George Schairer gathered in Germany. The swept wing allowed the B-47 to be significantly faster (550 mph in cruise) than the competition. The B-45, for example, cruised at 365 mph. The highly swept, high-aspect-ratio B-47 wings required the solving of a number of aeroelasticity issues so that vibrations, flutter, and aeroplane dynamics[9] were always under control. The plane's speed also forced understanding of airflow in the transonic flow regime, *i.e.,* close to the speed of sound. These were investigated on wind tunnel models and on the full-scale aeroplanes and were successfully integrated into a well-functioning, practical aeroplane design. Boeing's forethought of having had a wind tunnel with transonic speed capability turned out to be a key ingredient in allowing the company's aerodynamicists to master the design aspects in the new high-speed flight regime.

The bomber competition took place in 1947 with the B-45 entering service in 1948 and the B-47 doing so in 1951.

The high speed of the B-47 expanded the speed *range* over which the aeroplane had to be flown with sufficient lift, importantly at the lower take-off and landing speeds. This led to a long period of pioneering work at Boeing to develop high lift devices on the leading and trailing edges of the wings of their aeroplanes to an even greater capability than was required for flight near 300 mph.

---

8. A picture of Glenn L. Martin holding a model of the B-48 is shown in the personae section.
9. The Dutch roll, specifically.

120 *Powering the World's Airliners*

Fig. 47: Boeing B-47 (RB-47E) in flight. (Boeing Images BI210086).

The large B-47 turbojet-powered aeroplane unearthed some interesting aspects of operating it in the real world. The aerodynamics staff at Boeing strove and succeeded in designing a low-drag aeroplane. That allowed it to cruise fast, but it also brought about challenges that would be solved in this aeroplane and future airliners. The method used by people then and the current state-of-the-art will be seen to differ substantially.

The first operational challenge was the low thrust of the engines at take-off. The solution for the US Air Force was to incorporate a rocket assist for situations where it was needed: high take-off gross weight, high altitude airfields, hot days, or operation on short runways. The installation of the so-called RATO (Rocket Assisted Take Off) bottles was heavy if the installation was kept on the aeroplane. In a later version of the aeroplane design, the rocket array was mounted externally and jettisoned after use.

Some versions of the B-47 overcame the need for take-off thrust with water injection in the J47 engines that increased thrust by some twenty per cent but with a high cost in terms of smoke emissions. Water injection was used on the early turbojet engines with two effects: increased mass flow rate exiting the nozzle

Fig. 48: A rocket-assisted B-47 take-off and landing with drag chute. (Boeing Images BI 24698 and BI212308 respectively).

and an opportunity to increase fuel flow rate without incurring overheating of the hot section. The use of water injection or water/methanol mixtures saw practical service until the introduction of turbofans.

Another available alternative was use of engines with afterburning. This approach was and is common on military combat aircraft. None of the above solutions would be acceptable in the civil airport environment where noise and air quality constraints would be in place. In time, the supersonic Concorde airliner would be granted a singular exception to the noise rules that allowed use of afterburning on take-off.

On landing, the aeroplane had a different kind of problem. Its low drag, even with flaps deployed, resulted in a long runway roll. To compound the situation, the jet engine contributed thrust during a landing in order to be safe for a possible 'go around'. That thrust level was substantial because the engines were slow to accelerate to full power by virtue of the inertia of the rotating machinery and the need to control the power run-up while avoiding compressor surge or stall.

In landing, good practice is for the pilot to retract flaps on touchdown so that the wheels carry as much of the aeroplane weight as possible as soon as possible so that the application of brakes is effective. The previous generation of propeller driven aircraft had additional options. These included adjusting variable pitch propellers, and engine throttles set to produce as little thrust as practical. Some, the Lockheed Constellation for example, even had the option of reversing the propeller pitch to affect reverse thrust. This could be done because the piston engine was, relative to the jet engine, quick to respond to pilot demands for power setting changes. The solution for the B-47 was to deploy a parachute to produce drag on landing.

The rocket take-off assistance and the drag chute were solutions that could be used in the military setting, but their use in commercial operations would have been impractical. These were issues that the military planners in the Pentagon and at Wright Field debated as they contemplated the transition from piston-powered to jet-powered bombers. Fortunately, innovations on both the propulsion and aerodynamic sides of the aeroplane design came about to obviate the need for either of these rather spectacular approaches to dealing with the characteristics of the new jet aeroplanes. In time, the turbofan with thrust reversers and wing spoilers[10] would deal successfully with these issues.

---

10. Spoilers are wing upper surface panels that pivot upward hinged nearer the wing leading edge on the inboard side of the wings (aft of the line of maximum thickness). They are used for decreasing lift on descent or landing and for roll control in flight. Credit for their invention around 1948 is given to the Martin Company.

Fig. 49: Left: Wing lift spoilers deployed on an EasyJet Airbus A319 after landing touchdown. Note winglets at the wing tip. (Wikimedia Commons, John Haslam). Right: winglet details on an Alaska Airlines Boeing 737-900. Note vortex generators on fuselage between tail surfaces (Photo by author).

The Boeing B-47 had wings mounted high on the fuselage enabling the engines to be mounted below them. It was a beautiful design. It required the main landing gear to be mounted on the body of the aeroplane in a bicycle fashion with outrigger wheels at the inboard engine nacelles to stabilize the aeroplane on the ground. High wing designs were not to be used on Boeing's commercial aeroplanes. They used *dihedral*, *i.e.* sweep upward towards the wing tip instead. That allowed the use of more conventional tricycle landing gear wheel arrangements, while preserving the ability to mount the engine under and ahead of the wing. In that sense the B-47 was not a model for commercial aircraft to follow. The only large production jet aeroplane to follow the B-47 in terms of wing mounting design was the later B-52.

B-47 Legacy

It would be hard to argue against the notion that the B-47 changed everything. It did. The technology that went into its design came from a variety of quarters and the wartime development of a new engine type was central. The swept wing, the jet engine and its installation in podded nacelles together with wind tunnel capability allowed this aeroplane to be built. It put Boeing again in a leading position as an aeroplane builder and this time they would ride the wave into the future.

Not to be omitted from the tale is the General Electric Company that not only saw a future in the jet engine, but also realized from the early work with radial flow compressor engines that engine development could not proceed very far into the future. Thus they switched focus onto the axial flow compressor and built the

TG-180 (in 1943) that became (in 1945) the J35 manufactured by Allison and, later, the J47. The sleek form of this engine type made the design of the B-47 as beautifully practical as it was. The aeroplane also foretold of the demise of piston engines driving propellers, at least for large commercial aircraft. The life of the radial compressor gas turbine engine was indeed short, just about one decade. A similar length of time would characterize the transfer of the big piston engines into the history books.

Beyond industrial consequences in the United States and technical leaps generally, one could say that the B-47 was also a milestone in world affairs, specifically the evolution of the Cold War. The land-based bomber was then the central linchpin in the confrontation between the West and the Soviet Union. The arms race that followed the B-47 introduction was certainly a time when new ideas in aviation were tried on all sides, including the development of intercontinental ballistic missiles and ventures into space. Engineering education was a thriving business and valued by the nation. It was an interesting time.

Chapter 6

Jet Airliners

The Pioneers

As large jet powered aeroplanes became reality, innovative aeroplane builders took a serious look at using the jet engine to power an airliner. The first to put an aeroplane in service was the British firm de Havilland, which in 1952 delivered the first DH 106 Comet to British Overseas Airways (BOAC). Right behind it was the design house Tupolev in the Soviet Union that had built and put into service its Tu-104. This twin-jet airliner, with engines close to the fuselage and in the wing-root, went into service with Aeroflot in 1956. The Tu-104 might have been designed to serve a military purpose because it was adorned with a glass nose for a military (bombing?) mission, if its capability was required during the then 'warmish' Cold War. It initially seated 50 and evolved to a capacity of more than 110 passengers.

The Comet was a beautiful new way to travel if external appearances are an indication. Literature from 1950 shows an elegant aeroplane with coat- and hat-room and separate lavatories for gentlemen and ladies. These features did not survive to the present day.

In general, an aeroplane sized for the capability of turbojet engines at cruise conditions will have thrust levels that are wanting for take-off. The B-47 faced that problem with rocket thrusters to be used to keep the take-off roll sufficiently short. The Comet faced that same issue. The Comet-1 model prototype was equipped with rocket engines located between the two Ghost engines on each wing and their use was contemplated for take-off assistance. These were 5,000 lbs thrust de Havilland Sprite rockets fed with hydrogen peroxide as a monopropellant. The gas produced was environmentally benign oxygen and steam and no combustion was involved. The run time was about sixteen seconds. It was successfully tested on the Comet but never used in service because performance improvements of the Ghost engines obviated the need for them. On the military side (in the Vickers Valiant bomber), similar Sprite rocket modules were designed to provide the same function on heavy bombers with the additional feature that the rocket system package was jettisoned after use and descended by parachute to be recovered near the end of the runway. Mercifully, turbojet engines grew in capability and the later introduction of turbofans solved the take-off thrust problem once and for all. The rocket assist solution would probably not have been approved by civil aviation authorities.

Being a pioneer has its problems. The de Havilland was to experience the first of a number of design issues that were unearthed by virtue of being first. These included structural issues associated with vibrations, cyclic pressure loading of the fuselage (pressurization and depressurization on every flight), and stresses around the large rectangular windows. The figure opposite partially shows the windows on the prototype Comet-1. In part, the failure due to stress concentrations at the window corners led to a three deadly ('hull loss') accidents (two in 1954), making the history of that aeroplane one of production interruptions to correct whatever issues were discovered through the accidents.

The square corner window design on the early Comets was a logical evolutionary continuation of industry practice. The Boeing Model 307 Stratoliner, for example, had windows of a similar design and was not bitten by the consequences of this feature. It was designed to tolerate a 2.5 psi pressure difference allowing the 307 to cruise at 20,000 feet with a cabin pressure equivalent to an altitude of 8,000 feet. The Stratoliner experienced modest stress levels on the fuselage structure because it did not cruise as high as the Comet. The Comet was designed to cruise at 42,000 feet, where the atmospheric pressure is quite low. If the cabin pressure were held at 5,000 feet, a typical norm at that time, a pressure difference of about 10 psi would be experienced by the fuselage. The Comet's exposure to stress was about four times more severe.

After the accidents and discovery of the structural failure in the wreckage, the engineers addressed the problem. They carried out experiments involving a full-scale production fuselage filled with water and placed it in a large water tank. That way the fuselage could be easily and safely pressurized over many cycles until the failure was replicated. And so it was that a redesign, including oval windows, led to a better product.

To the good was that the Comet accidents initiated the thorough methodology employed thereafter by safety agencies the world over in examining all flight-related accidents, including the later use of flight data and cockpit voice recorders.

The Comet was initially a small aeroplane with a take-off gross weight of 105,000 lbs. The Comet-1 model had 36–44 seats in 2+2 seating, and it was fast, 460 mph in cruise. It was aerodynamically clean with a relatively modest wing-sweep and four engines buried in the wing-roots. The wing-root location of the engine inlets was implicated in one accident involving wing stall. The issue was thought to have been engine compressor stall due to lack of 'clean' air when the wing stalled.

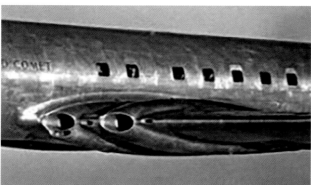

Fig. 50: The de Havilland Comet-1 prototype flying in 1949. (Courtesy of BAE Systems plc, Neg. No. DH4392). The picture at left shows the square windows and the inlets of the -1 model.

The inlets were subsequently modified. From an airline's viewpoint, this kind of engine location also presented operators with a risk of airframe damage, the wing in particular, in case of mechanical engine failure or engine fire. Fortunately, these issues were never experienced.

The Comet-1 had four de Havilland Ghost 50 Mk1 radial-flow compressor engines each producing 5,000 lbs of thrust. The Comet grew in size over the years

Fig. 51: Interior of a de Havilland Comet 4C in the restoration shop of the Museum of Flight. The original 2+2 seating in the front and the addition of a fifth seat per row are shown. The aeroplane had generous legroom in all rows and round windows with curtains. Note the absence of overhead storage bins for luggage. (Photo by author at the Museum of Flight).

it was produced until 1964 when production ended. Rolls-Royce Avon[1] engines were used, starting with the Comet-2 model. These engines were developed with increasing thrust levels from 7,000 lbs to over 10,000 lbs. The last of the Comet series (-4) was introduced in 1958. It weighed in at 156,000 lbs and accommodated over 80 passengers with 2+3 seating possible, though perhaps not comfortable. Later versions of the Comet also featured what is today an uncommon approach to storage of sufficient fuel for long-range routes: external wing-tanks at mid-wing. The number of Comets built was 114. The Comet was a worthy pioneer and paid the price of insufficient capability compounded by a lack of design experience that led to accidents and a jumpy market for their product. It showed the way to the future and taught the builders to follow what details must be meticulously addressed.

---

1. The Rolls-Royce Avon was that company's first axial flow compressor engine.

While the Comet was grounded, the Tu-104 was the only jet airliner in service. About 200 Tu-104s were built between 1956 and 1960 and the aeroplane was retired from service in 1986. The Tupolev firm went on to design and build a number of follow-on airliners (Tu-124 and 134), most of which were sold to national airlines behind the Iron Curtain. These airliners grew in capability as technology evolved, but purchase interest in the West was slim to non-existent. The most memorable of the Tupolev designs was the supersonic Tu-144 that was alleged to be the product of industrial espionage aimed at the Anglo-French Concorde. The Tu-144 never saw serious service and ultimately suffered a bitter end.

The British Comet and the Avro of Canada Model C-102 first flew in 1949. The C-102 was built as a prototype and never went into production. It was similar to the Comet except that the engines were further out on the wing. Initially, the C-102 design was tailored to the specifications of Trans Canada Airlines (TCA) with Rolls-Royce Avon AJ.65 engines (an axial-flow compressor design with a thrust of 6,500 lbs) that were then not available for civilian use on account of military classification. When the ability to use these engines failed to materialize, the design was modified to use radial flow compressor engines (3,600-lb thrust Rolls-Royce Derwents) mounted in pairs on the wing though not podded as on the B-47. The engine substitution was a setback for TCA because the axial-flow engine installation was much more attractive from the viewpoints of aeroplane cruise speed and payload capacity as well as 'beauty' of the nacelle. Additionally, the redesign involved development costs and delays in delivery schedule to the point where TCA lost interest. Further sales efforts to US airlines by Avro of Canada were not fruitful.

The B-52 and Dash 80

In the years that followed the successful deployment of the B-47 bomber in the US, groundwork was being laid for an even bigger aeroplane destined to become the B-52. American jet powered airliners were also becoming a possibility. At this time, the planned deployment of the B-52 would require a tanker faster than the KC-97 then in use. This piston-powered tanker was a derivative of the Boeing Model 377 Stratocruiser. The speed mismatch of the KC-97 and the new B-52 mandated a keen US Air Force interest in a jet powered air-to-air refuelling tanker.

These factors and the experience in hand led Boeing to develop a transport aeroplane that could be used as a military tanker or a civilian transport. This was the Model 367-80, commonly referred to as the Dash 80. The model number

130   Powering the World's Airliners

Fig. 52: Boeing Dash 80 with an Air Force KC-97 aerial tanker at Renton Field on the shore of Lake Washington. The KC-97 is the military version of Model 377. (Boeing Images BI29584).

367 was in fact a version of an aeroplane based on an existing airframe with four engines on a straight wing. Rather than assign a new model number to the new jet powered aeroplane and thereby reveal, perhaps too widely, that Boeing was designing a swept wing jet aeroplane, it was decided to number this aeroplane as Model 367 with increasing version numbers as the configuration took final shape to become the 367-80. Construction was authorized in 1952 and it was rolled out in 1954. This largely hand-built prototype had a body diameter of 132 inches, sufficient for 2+3 seating. It employed a swept and low wing with podded engines. The engine installation was somewhat similar to those of the B-47. These design elements would end up in many future transonic transports. By the time the Dash 80 was designed, available engine thrust increases allowed use of four engines for the design weight of this aeroplane.

Pratt & Whitney and the J57

The details surrounding the process by which an engineering and engine construction business evolved to develop a new product is not often well documented. Thus, the stories of people and their thinking are seldom available to shed light on how important steps in technical innovation are realized. One such document was a recorded discussion held at the New England Air Museum concerning the development of the Pratt & Whitney J57 turbojet. The recording dates back to the first decade of the 21st century and preserves the essence of the events at Pratt & Whitney in the early 1950s. The authors were J. Connors, T. Slaiby, and H. Schmidt, retelling of their close involvements with the building of this important engine. This story necessarily involves some technical details because it is these details that are the totality of the breakthrough that this engine was. These details will be described below in general terms.

It is the time of the Cold War in the years immediately following the end of the Second World War. The United States' confrontation with the Soviet Union and the arms race were in full swing. The goal of the air-side of the American military establishment is to have in service, among other things, long range bombers capable of reaching the Soviet Union. The ten-engine B-36 was designed to be able to do that but it was born in the waning days of piston engine technology and was obsolete by the time it was built. The jet age had dawned and the medium range B-47 bomber was operational. The B-36 needed to be replaced by a more capable aeroplane. It was clear that the new heavy bomber was going to have gas turbine propulsion. The dilemma for aeroplane designers was that new powerful turboprop engines would be available and these would allow for the achievement of the necessary range. But a turboprop aeroplane would be relatively slow. On the other hand, a turbojet-powered aeroplane would be fast but the technology available would limit the aeroplane's range.

The Boeing Company's investigations into an aeroplane not yet known as a B-52 involved a straight-winged turboprop Model 462 and a later turbojet-powered Model 464. As far as the US Air Force was concerned, turboprop engines were to be developed by Wright Aeronautical and the use of the Westinghouse J40 was considered for the turbojet-powered aeroplane. Pratt & Whitney followed the evolving propulsion need with interest. The company also needed a product to surpass the good position enjoyed by General Electric as they had successfully built the engines for the B-47 and other military aircraft.

The performance of a gas turbine engine as a thrust producer in that period was limited, in part, by the capability of the compressor. Junkers and Whittle had started with compression ratios of three for engines used during the war and, in a few years, that parameter had risen to about six in the axial-flow J47. The general consensus was that higher compressor pressure ratios would be necessary but would be difficult because internal air-leakage would be hard to control.

This is the challenge that Pratt & Whitney took on as they looked at engine configurations. A turbojet with a higher compressor pressure ratio would satisfy the speed and come closer to meeting the range requirements. Their approach was to configure the compressor in two parts, in series. A first compressor with a pressure ratio of about three would be followed by another with a ratio of four, giving a total of twelve. The design details involved in such a *two-spool* engine are described further on, but for now, a number of the necessary innovations for a successful engine with this significantly higher pressure ratio are worthy of mention. The first is that the engine would be longer than the engines of the day. The shaft connecting the compressor and turbine is therefore also longer and more prone to buckling at high rotation speeds unless it is made stiffer and therefore heavier. One part of the solution was to shorten the length of the combustion chambers. Indeed, Pratt & Whitney succeeded in the design of the so-called *can-annular* combustion chamber that ended up in common use. These combustors had a clever inner 'cold' air feed at the centre of the so-called 'can'. Figure 53 illustrates this innovation on the lower of the two cans shown.

Another problem that had to be solved was that the blading near the exit of a high-compression-ratio compressor would be very small and delicate if the compressor had a cylindrical outer case and a narrowing inner case, a commonly considered way of building such a device. In such a configuration, the flow at the outlet would be dominated by friction effects and the efficiency would be

Fig. 53: A ¼ scale sectioned model of a P&W JT3 (J57) showing the low- and high-pressure components. The model has 9 LPC, 7 HPC, 1 HPT, and 2 LPT stages. Note the diffuser at combustor entry. (Picture taken at the National Air and Space Museum, Washington, D.C. by Sanjay Acharya.)

Fig. 54: A cutaway view of a Wright J65 turbojet (ca. 1952) showing how (according to Professor Söderberg) not to build an axial flow compressor (with a cylindrical outer case). (Wikipedia Commons).

poor. Professor C.R. Söderberg of the Department of Mechanical Engineering at Massachusetts Institute of Technology suggested that the compressor case be made to decrease in diameter in the flow direction to keep the blades relatively long. Good idea, but the flow conditions are more complicated and very sophisticated design procedures would have to be devised to keep the pressure rise as uniform as possible. Further, longer blades are also more prone to vibrate. These challenges were indeed overcome thanks to good engineering analysis, although it was probably not done solely at Pratt & Whitney! The resulting engine had a wasp-like waist appearance and, more importantly, it became lighter.

Finally, an innovation developed was the use of a blade end shroud[2] on the turbine rotors to prevent high-pressure combustion gas from bypassing the turbine and thus robbing the turbine of getting work out of any leakage. The shroud connected the blade tips mechanically and, as an added benefit, reduced blade vibration. The use of blade end shroud was common in the steam turbine industry, novel in a jet engine, and to this day not universal in the latter application. For example, the T53 engine shown in Fig. 71 (see page 174) has an unshrouded

---

2. The idea of the shroud on a gas turbine blade may be novel here but the invention dates back earlier (probably in connection with steam turbines). See for example US Patent 2278041A 'Turbine Blade Shroud', by R.C. Allen (Allis-Chalmers) granted in 1942.

high-pressure turbine with cooling-air exiting the blade tip and a shrouded power turbine.

The decision to double the compression ratio over then common practice was not simple and impacted a number of design areas. These were all successfully dealt with under the far-seeing leadership of Leonard Hobbs, engineering vice president. For this he won the Collier Trophy in 1952. In retrospect, it can be said that development of the J57 was a genuine tour de force for Pratt & Whitney. It was chosen to power the B-52 and the engine bore the in-house and civilian name of JT3.

The B-52 first flew in April 1952 and since 1955 has served with the US Air Force. Its important design feature was the achievement of speed allowed by the use of turbojet engines. This design decision impacted the range of the aeroplane and that limitation led to thinking about using in-flight refuelling to extend the range of the B-52. Naturally, since the B-52 was going to be travelling fast, it made sense to have an air tanker that was also fast, faster than the Stratocruiser-based design of the KC-97 then in operational service. The J57 went on to become Boeing's choice of engines for not only the Dash 80 intended for service as a tanker, but also for the later 707.

The J57 was also used in military fighter aircraft of the day, specifically the North American Aviation F-100 Super Sabre (Fig. 70, see page 172) and the McDonnell F-101 Voodoo. Much was learned from the much more demanding operating conditions of a fighter. At that time, a very important challenge was compressor stall and how to avoid it. The rather benign flying conditions experienced by a bomber or a transport tended not to reveal as many problems with stall as did the severe high-g turns and steep ascents of fighter aircraft. The J57 benefitted from its use in fighter aircraft and was well placed in the large aircraft.

History notes that the Soviet Union took the other route for the long-range bomber: the Tu-95 bomber is illustrated in Fig. 41 (page 101) with its four counter-rotating propeller turboprop engines delivering almost 15,000 HP each.

The transformation of Pratt & Whitney from a leading manufacturer of air-cooled piston engines to a position where they were able to design and manufacture a very good jet engine is quite remarkable. It took place relatively quickly. The experience of subcontracting production of the Westinghouse J30 engines and the Rolls-Royce licensed J42 and J48 surely initiated the process, but engineers and machinists had to be involved who, very likely, were not trained or experienced in this new field and had to adapt quickly.

The 707 and DC-8

The appearance of a commercial version of the Dash 80 available for sale to the airlines lit interest in others, notably Douglas, who rose to the opportunity to design a long-range jet of their own. Competitive pressure from Douglas with a proposed DC-8 ultimately led to the production version of the Dash 80, the 707 (a model number invented for marketing purposes), to have a larger fuselage diameter of 148 inches, sufficient for 189 seats in mixed class, 3+3 seating. The 707 production aircraft included a new improvement to the high lift system with installation of Kruger flaps on the wing leading edge between the engines. The airline customer who launched production of the 707 in the fall of 1955 was Pan American Airways, the same airline that first ordered the Boeing 314 Clipper before the war. Service with this new long-range airliner began in 1958. Pan Am was also the launch customer for the Douglas DC-8 that entered service in 1959. The Pratt & Whitney JT3C turbojets powered the 707 as well as the DC-8. The Pan Am orders were for twenty 707s and twenty-five DC-8s. No major airline wanted to be left behind and orders for the Boeing and the Douglas aeroplanes were placed in quick succession. TWA was among them with a small order of eight 707s. By the end of 1955, orders were in place for 186 jet airliners. This period essentially determined who the main aeroplane builders were going to be, but might there be room for a third?

Convair, Hughes and TWA

The following paragraphs are gleaned from the much larger story well told by Robert Rummel in *Howard Hughes and TWA*, cited earlier. Hughes played a very interesting, one could say disruptive, role during those early years of the jetliner industry that included the airlines as well as the aeroplane and engine builders.

When the initial deluge of orders for the 707 and DC-8 came through, Howard Hughes was a majority shareholder in TWA and keenly interested in playing a role bigger than buying aircraft from Boeing or Douglas. Hughes thought of paralleling the advantageous position he devised for TWA in purchasing the original Constellations. The plan was to involve the expertise and facilities of an existing airframe manufacturer and get the Hughes Tool Company to become an alternative to Boeing and Douglas. Such a plan required approval from the Civil Aeronautics Board (CAB). He proposed modification of an earlier understanding he had with the CAB that had given him a role in the building of the initial set of Constellations. The CAB however rejected his proposal and that approach came to naught.

Fig. 55: One of the TWA domestic service 707s. Note the daisy-petal noise suppression nozzles on the outboard engine. (Photo Jon Proctor).

Here the story of aeroplanes and engines becomes even more interesting. The order of 707s for TWA was increased from 8 to 33 in the following year, January 1957. This order, consisting of domestic 707-131 and international 707-331s,[3] was not, however, from TWA but from the Hughes Tool Company (Toolco) and there was no understanding that the aeroplanes were intended for TWA! In fact in January 1957 Hughes told Rummel, 'TWA has no rights whatsoever to the Toolco jets.' The thinking might have been that he retained control of the aeroplanes until *his* aeroplanes were available for TWA. Hughes wanted to build aeroplanes and the Boeing aeroplanes were to tide TWA over until Hughes could control the manufacture of his new aeroplanes and enter the market.

Initially, Hughes' options included 1. getting Boeing to consider designing a smaller 707, 2. using the Lockheed Electra turboprop, or 3. working with Convair to build a smaller all-jet aeroplane, to his specifications. Boeing was not (yet) interested in starting to design a new aeroplane, even if it had commonality with the 707. The order book was full, the factory was busy, and the first delivery was still in the future. It seems reasonable to speculate that if Hughes had talked Boeing into building a smaller 707, he would have tried, probably unsuccessfully,

---

3. In this time period, domestic aeroplanes were designed for a range of about 3,500 miles (US coast to coast with reserves) while the aeroplanes for international routes had to be able to fly about 1,500 miles further.

to control deliveries so as to exclude TWA competitors from access to them. The Electra option was judged to be too short a step into the future. Choosing the last option of working with Convair was certainly closest to his heart.

The question as to whether a third airliner builder was to be in the jet business came closer to resolution in 1958/9 as the 707s and the DC-8s were being delivered to airlines and put into service. Thus Hughes played the Convair card with an order for thirty aircraft that became the CV-880. Delta Airlines augmented this order with ten aeroplanes for its fleet. One stipulation of the purchase contract was that production of CV-880s was to be limited to TWA and Delta Airlines, at least initially.

Thus, Convair (a division of General Dynamics Corp.) joined the fray with introduction of the Model 880. It was similar in design layout to the 707 and DC-8, albeit smaller (110 passengers, 2+3 seating) and claimed to be faster. Its design cruise Mach number was said to be 0.91, faster by about 35 mph than the 707 and the DC-8 with the attendant bragging rights. The latter two cruised economically at about Mach 0.83 to 0.85.

The Convair 880 was powered by four General Electric CJ805 turbojets. These engines were commercial derivatives of the military J79. Unlike the twin spool Pratt & Whitney J57, the J79 was a single spool design. The J79 had been developed for military aircraft of various kinds, including the Convair B-58, the Lockheed F-104, and the McDonnell F-4 (Phantom II). These were all successful supersonic aircraft. The B-58 was the replacement for the pioneer B-47 starting in 1960. The CJ805 engine was GE's 1956 entry into the commercial market and its fate would be tied to the bumpy journey of the Convair 880 and its follow-on, the 990.

Hughes had failed to make an early decision to buy 707s and started the CV-880 construction programme, presumably to catch up. He was now feeling the competitive heat as TWA's competitors – Pan Am, United, and American Airlines – enjoyed a pleasant reaction to the jets introduced to the travelling public. At one time he tried to get a better delivery position for his 707s with an attempt to litigate for correction of an imaginary understanding he had with Boeing about delivery position in the assembly line. He hated seeing TWA late in the adoption of the new 707. That did not go well for Hughes because Boeing had a firm hand on the tiller and a strong commitment to customers to maintain an orderly delivery schedule. The first Toolco aeroplane was the eighteenth off the assembly line.

Hughes was negotiating for more than the 707s already on order, when Pratt & Whitney offered to sell JT3 and JT4 engines to airlines, presumably for spares. The JT4 was an advanced, more powerful engine similar to the JT3. He immediately opted to buy 210 engines (thirty JT3s and the rest JT4s) for earliest delivery! These numbers were far more than were needed for the sale being negotiated and caused Pratt & Whitney to hesitate about proceeding. It foresaw difficulties that this order would place on future sales of the engines to other airline customers, including Boeing and Douglas. Pratt & Whitney hesitated and Hughes raised the order to 300 engines. Pratt & Whitney caved in and accepted it. Rummel considered the transaction to border on the unbelievable because a fleet of twenty 707s foreseen at the time of the negotiations needed no more than sixty engines as spares. Rummel's relatively large estimate of the number of spare jet engines needed reflected his experience with piston engines; the smaller need for jet engine spares was not yet apparent in the airline industry. He speculated in his book that Hughes might have been thinking about dreams of supplying Boeing or other builders with engines whose price and availability he would control. The play was an attempt at cornering the engine market for the aeroplanes that would compete with his Convairs, pure and simple. Fortunately for the industry, when cash flow problems later arose at Toolco, the engines were successfully sold out of his control with the assistance of Pratt & Whitney.

Hughes' financial circumstances also prevented him from meeting the payments required when the sixty-three (33 from Boeing and 30 from Convair) aeroplanes ordered for Toolco were ready. As a result, he reduced the order for the 707s. Boeing was well able to accommodate this request because there were customers happy to jump ahead in the delivery schedule. In fact, six of the long-range 707s were sold to TWA's rival, Pan Am, with the necessary approval of Hughes. Juan Trippe, chief executive of Pan Am, was very pleasantly surprised for this astounding turn of events. He got his rival's aeroplanes!

The Convair 880 order was reduced from 30 to 20 aeroplanes and the matter was much more difficult in that arena. Toolco, not TWA, was the main customer for the aeroplane. Hughes's inability or unwillingness to make timely decisions about a myriad of details led to production delays. Whether these delays were imposed on Convair to avoid the necessity of accepting and paying for the airliners as they were completed is debatable. Faced with the uncertainty of whether and when the Convair jets were to be made available, TWA had an excruciatingly hard time planning for the integration of a yet-to-be-determined number of jets into its fleet.

Delta was taking delivery of their ten 880s interspersed in the production line and was undertaking flight tests in anticipation of fleet use when one of the Delta CV-880s crashed on take-off, killing the crew of four. The cause was never finally determined, although a number of improvements were made as a consequence of the accident.

Hughes, for his part, still did not allow his aeroplanes to be finished and accepted. For Convair, the situation could not have been worse. Unfinished aeroplanes were stored at the plant and there were no customers to whom they might be sold. Convair was at the mercy of the uncertainty sown by Hughes. In spite of recommendations for a solution to the financial crisis from TWA management and his own staff, he did not allow a rescue plan to be developed that included the investment community. The banks would surely want some leverage to protect their interest in any financial rescue plan. Hughes was unwilling to give up control of any of his assets.

TWA did manage to obtain access to one of the 880s, in order to get ready for fleet integration, if and when that was to occur. The first flight from San Diego to Kansas City revealed that fuel consumption was significantly worse than anticipated. This sounded alarm bells at TWA because some projected routes that were marginal from a profit viewpoint would become loss generators.

On the last day of 1960, Hughes' financial difficulties climaxed with the establishment of a plan to solve the mess that had been the relationship between Hughes and TWA. Hughes lost his voting rights to an investment trust set up by the company management. That transaction allowed the procurement of aeroplanes to become orderly and TWA to actually receive the 880s ordered by Toolco. Matters improved for TWA and they successfully introduced them into service in early 1961.

For a number of reasons however, the 880 programme at Convair was looking like a failure, and steps were underway to produce a better aeroplane that became the CV-990. The 880 production line continued with deliveries to a variety of airlines until 1962 with a total of sixty-five aeroplanes. The fuel consumption and the maintenance costs of the CJ805s were part of difficulties suffered by the operators of the CV-880. Making things worse for Convair was the Boeing offering (in late 1957) of a Model 720 also aimed at the medium range market. The 720 was a smaller variant of the 707. The TWA assessment of the 880s is that the airframes were excellent, but the engines were less so. In his book, Rummel indicated that TWA had specified changes to the engines that had made them smoky on take-off. TWA probably desired greater take-off thrust by water injection. The airport vicinity community was not happy. In the end, TWA retired the 880s early in 1973.

The involvement of Howard Hughes in aeroplanes and engines ends with the CV-990 and the GE turbofans to be discussed in the next chapter, but the reader would be mistaken to expect that the litigation for his attempts at a recapture of TWA ends here. It went on much longer.

Howard Hughes, a postscript

No story of the American airline business is complete without the inclusion of Howard Hughes. He was a man of many interests and wealthy enough to really get into his love of aviation. That aspect of his enigmatic story ranges from air racing in his younger years to airline ownership in the age of the big piston engine aircraft, and into the jet age. His aggressive passion gave the world new aeroplanes, airliners, and turbulence on a grand scale. His business goal was exclusivity in the airline business and an obsession to be a leading figure. He had a significant impact on the evolution of the jet engine business in the United States as described above and more to come below. He was a complicated person.

He tried to stay in the airline business with the acquisition of Northeast Airlines in 1960. It was resold to a trustee shortly thereafter. In 1970 he purchased Air West and renamed it Hughes AirWest. The airline operated for a decade and was sold to Republic Airlines in 1980 after Hughes' death in 1976. Its assets were merged into other corporate entities.

He is probably best known for his construction of the all-wood mega aeroplane known as the H-4 Hercules. This mostly birch plywood flying boat, commonly known as the *Spruce Goose*, was initially designed to cross the Atlantic during the Second World War when shipping by boat was so threatened by German U-Boats. The reason for the wooden construction was centred on the need to build aircraft without using critical materials and resources, such as the electric power required to make aluminum. Due to a dispute with the government, the construction of the H-4 was limited to just one aeroplane that was not completed in time to play a useful role in the war. It was flown only once, on 2 November 1947, to demonstrate that it could fly. The contractual obligation to the government was thereby met and Hughes was paid.

Robert Rummel describes Hughes as he interacted with the officials at TWA. The reader of his book cannot escape the conclusion that Hughes was obsessed with privacy and secrecy. Rummel was trusted by Hughes and was the channel to TWA management, but even he could not contact Hughes on his terms. He had to leave word with Hughes' office that contact was desired and await a return call, sometimes for days or weeks and then usually at times an ordinary person might not call 'business' hours. While the manifestations of Hughes' decisions regarding TWA are well documented in the book, the reader may wish to learn more about the man. Much is written and some may never be well understood.

Turbojet Engine Family Trees

The American military establishment played a major role in bringing forth the turbojet engines that ultimately found their place on commercial airliners. They needed to find the best way to build them for the Cold War arms race. The two major American engine builders, GE and Pratt & Whitney, each developed competitive lines of military turbojet engines whose technology would be incorporated in the commercial engines. The historical connection between Rolls-Royce and GE also led Rolls-Royce to be a strong part of the evolution.

Rolls-Royce
The early history of British engines started with radial flow compressor engines. These engines were the Welland, the Derwent, the Nene, and the Tay, in that order. These names follow a Rolls-Royce tradition of naming most gas turbine aircraft engines after rivers in the United Kingdom. The technology of these engines had roots in both the UK and the US since these countries cooperated closely. For example, the Nene was a Pratt & Whitney J42 (see Fig. 35, page 91) and the Tay was also produced as a Pratt & Whitney J48. These two Pratt & Whitney engines powered many of the US Navy's Grumman military fighter jets during the 1950s and were used in the Korean conflict.

In that time frame, Rolls-Royce switched to the axial compressor with its design of the Avon engine. This engine went through a naming sequence that should lead to head shaking, but here it is: the AJ.65 (Axial Jet of 6,500 lbs of thrust) entered production (in 1950) as the RA.3 with various version numbers, and finally acquired a longer-lived standard Rolls-Royce naming as the RB.146. From an engine man's viewpoint, the Avon was a cohort of the Pratt & Whitney J57, the Allison (GE) J35, and the SNECMA Atar engines. There were differences of course, but these turbojets set the stage for all developments to come. They all were produced in great numbers with the production of the Avon running from 1950 to 1974.

General Electric
GE's initial success was with radial compressor engines that were largely produced by other companies, notably Allison. GE's early recognition of the virtues of axial flow compression resulted in the large production of the military J35. The company went on to design and produce an American pioneer for the jet age, the J47 (1948 to 1956), importantly for the B-47 and other aeroplanes. This engine morphed into the J73 and, later, the J79 that entered partially successful commercial service as the CJ805 and the aft-fan version, the CJ805-23.

Westinghouse

Westinghouse was an early proponent of axial flow compression engines. The company produced two such jet engines, primarily for the US Navy, the J30 and the J34. A J40 engine was also produced but was judged a failure. The company exited the aviation gas turbine business in 1960.

Pratt & Whitney

The success in the role of licensee producing the J30, J42 and later J48 engines led to Pratt & Whitney's design for their first significant axial flow turbojet engine. This was to be the J57 intended for the B-52. These engines were produced between 1956 and 1965. The company name for the J57 was JT3. It was widely used commercially in the early 707s and DC-8s until the advent of the turbofan when that engine was modified to become a turbofan, the JT3D. That engine became military hardware with its own appellation of TF33. The J57 was later developed into the J75.

The civilian version of the J75 had the in-house name of JT4 and was used on the 707 and the DC-8 for better take-off performance on the long-range models. It was about forty percent larger in airflow rate with a corresponding increase in thrust. The JT4 was not developed into a turbofan.

A scaled down version, about twenty per cent smaller, of the J57 became the J52 intended for a number of military applications. Pratt & Whitney's company name for this engine was JT8A. This engine would live long and successfully in many airliners when it was modified into a turbofan.

Since the 1990s, GE, Pratt & Whitney, and Rolls-Royce developed their turbojet engines further, but the development focus for commercial aeroplanes was the turbofans.

Turbojet engines as such did not see applications in commercial service after the start of the era of the turbofan. One exception is the Rolls-Royce/SNECMA Olympus 593 Mark 610 turbojet (31,000 lbs thrust dry, 38,000 with afterburning) used on the Concorde between 1976 and 2003. A turbojet engine named the GE4 was proposed for the Boeing Supersonic Transport (the ill-fated Model 2707, SST) as a derivative of a General Electric YJ93. It was designed to produce 50,000 lbs of thrust, 65,000 lbs with afterburning.[4]

---

4. Afterburning is the process of adding fuel to the exhaust jet to raise its speed by raising its temperature (*i.e.*, its sound speed) at the same Mach number.

Allison

After a heavy production period in the Second World War of the V-1710 piston engines, Allison Engine Company filled its production capacity with jet engines. It produced J33 (radial) and J35 (axial) jet engines for General Electric because of GE's limited production capability. A significant success for its own designs was the T56, a turboprop that was used in a number of regional airliners and the military C-130. In 1995, the Allison Division of General Motors was acquired by Rolls-Royce.

Jet Engine Improvements, Step by Step

From an aerodynamic viewpoint, the design of the compressor is the most challenging aspect of building a turbojet engine. Its evolution is marked by a number of important breakthroughs. The compressor is central to the design of the jet engine because it determines its power and efficiency as a thrust producer. The Whittle, Junkers, and the early GE engines had low compressor pressure ratios in their engines. GE and Pratt & Whitney raised that parameter in the J47 and the J57. There had been much room for improvement on that aspect of the design. From a technical viewpoint, there is a best pressure ratio for fuel economy and another for high thrust. In making the choice, one has to consider the temperature capability of the turbine and how well the compressor and turbine can perform their tasks. Thus, as the latter improved, so did the need for higher compressor pressure ratios and ingenuity in how the compressor was built.

Controls

Unlike the piston engine, the gas turbine or jet engine operates most efficiently at high power output so that its design can be optimized for operating near maximum output, but it must also be controllable to provide lesser amounts of power reliably. The sensitivity of airflow through the compressor blading, which must be oriented within a narrow range of angles, is the Achilles heel of the idea of the axial flow compressor. This challenging aspect of control was solved in the Jumo 004 with the movable centerbody in the propulsion nozzle. That solution was subsequently deemed impractical for modern engines and designers turned to fuel control systems exclusively to vary power output. Thus, postwar American engines like the J47 and J57 incorporated little or no variable engine geometry save for valves to allow starting the compressor. The fuel control system sensed the airflow rate and the rotational speed of the blading and took such measurements into account for modulating the demands of the pilot for power changes. The goal was to maintain efficient and reliable compressor operation not only in a new desired state but in the process of approaching it. Such power adjustments were slower than similar changes requested from piston engines and thus required pilots to adjust to a 'less responsive' jet engine. The simplicity of the early engines would later be compromised for better performance.

Spools

The earliest jet engines were of a one-shaft or one-spool engine design. The axial flow engines had fixed blades, meaning that the blades in the compressor were fixed in orientation. This will be improved in time for *stator* blades, although variable *rotor* blading remains impractical for the foreseeable future. Examples of these early fixed compressor geometry engines include the Jumo 004 and the GE J47 first run in 1947. The turbine simply drives the compressor as it was built. A later model of the J47, called the J73, introduced the variable inlet guide vane.

During 1945/6 the Bristol Engine Division studied engine configurations[5] for increased thrust and reduced fuel consumption and came up with the idea of using a *two-spool* engine configuration where the compressor and the turbine are broken into low- and high-pressure components. The idea was first run in 1950 in a British Bristol Olympus engine.

A two-spool Pratt & Whitney J57 (or JT3) turbojet also first ran in 1950. The idea of a two-spool compressor was the leapfrogging key for Pratt & Whitney to effectively get into the jet engine business. Whether the idea was spawned at Bristol, Pratt & Whitney or elsewhere is unclear but it was a very good one. The breakup of the compressor into two components is desirable because flow conditions at the front are quite different from those at the rear of the compressor. In a two-spool engine, a high-pressure turbine or HPT (please excuse the acronyms!) uses the combustion gas to drive a high-pressure compressor, or HPC. A low-pressure turbine, LPT, running on its own shaft behind the HPT passes its power by means of a shaft through the centre of the high-pressure engine shaft. It provides power to (yes, you guessed it…) the LPC. A two-spool engine of this type can be thought of as an engine within an engine. The LP shaft of the J57 typically runs at 5,800 RPM while the HP shaft turns at 9,600 RPM. The chief advantage of this configuration is that the LPC and the HPC can operate separately with better efficiencies. This results in a higher overall compression efficiency. With two spools, the engine's fuel consumption rate is lower than with a single shaft engine over a wider range of operating conditions. Looking to the future, we note that Rolls-Royce's large

---

5. Flight Global Archive, December 1955.

turbofans are alone in building the engines with *three* shafts running inside one another!

Variable blade geometry

It is noteworthy that in the earliest engines, the flow conditions, the flow orientation relative to the blades and the mass flow rate, into the compressor were determined by inlet guide vanes. These vanes are the stationary blades that air encounters on initial entry into the engine. Some later engines featured inlet guide vanes that were adjustable by a control system to control the engine's airflow rate, much like the 'onion' achieves the same task in the Jumo 004.

Airflow modulation was also carried out in the J47 engines like those mounted in the Boeing B-47. When water injection (for take-off) was used, airflow modulation was required. To that end, a small blocking vane was deployed into the flow of the exhaust nozzle of the engines. In normal operation the vane was retracted. That technique was not common and is no longer in use.

Fig. 56: The compressor of the GE4 engine destined for the ill-fated Boeing Model 2707 Supersonic Transport. This 9-stage compressor has variable geometry stators whose angles are varied by rotation of actuator rings tied to levers. The combustion zone is to the right. (Photo by author at General Electric, Evendale, Ohio).

A compressor in the early axial flow engines with fixed blading angles, both one- and two-spool designs, works well enough but it penalizes performance at conditions other than those for which operation is specifically designed. Fortunately, the long-range aeroplanes largely, though not exclusively, operate at a single operating point near maximum power or efficiency. The 1950s saw an important breakthrough in the control of the jet engine compressor with the filing in 1955 for a patent on a control scheme for varying the angle of the stator blades (and sometimes the inlet guide vanes) on the case of a compressor by a ring that turns all the stator angles of a stage in unison. The idea is to have the angle of attack on all blades of a stage always to be configured to maximize compressor efficiency and avoid stall. This approach is now universally used in gas turbine engines. The originator was Gerhard Neumann,[6] the renowned German-American gas turbine engineer at General Electric.

These variations in the design of compressors become all the more necessary in turbofan engines where the fan is normally running with the low-pressure spool. The fan introduces an even greater difference in operating conditions at the front and near the exit of the compressor, making all the improvements described above quite necessary.

---

6. 'Variable Stator Compressor' US Patent 2933235A, Granted to G. Neumann, 19 April 1960.

Chapter 7

The Turbofan Engine

The performance benefits of the turbofan were recognized early as a variant of the pure turbojet that the industry started with. Frank Whittle suggested as much in a patent application[1] in March 1936. He and others recognized that it is much better to obtain thrust from a jet engine by having a large air flow rate and a relatively small jet velocity. As initially built, the turbojet engine processed a *small* airflow with a *high* jet exhaust speed. The argument then goes, why not take more jet energy out of the hot gas by means of the turbine and feed the power to a fan to thus achieve a higher propulsive flow mass? The fan and the engine flows could be made to exit the engine out a common nozzle or out of separate nozzles. In practice, both options were common. The higher air mass flow rate raises the propulsion efficiency, an attribute that the propeller as a mechanical jet producer enjoyed with its *low* jet velocity and very *high air*flow rate. The result was significantly better fuel consumption on the part of the gas turbine engines. Whittle let that patent expire in later years.

Introduction of Turbofans

Engine technology was changing as the turbofan idea was revisited. In 1960 a version of the 707 was built with Rolls-Royce Conway engines, leading the industry in the use of bypass or turbofan engines.

Current fans in turbofan engines have rather low compression ratios (together with high bypass) as demanded by the need for low noise in the airport environment and good fuel economy. It is interesting to note that the Conway design took quite a different approach to inclusion of a fan. It used a portion of the low-pressure compressor output as the 'fan' or bypass flow. This low-pressure compressor had seven stages that resulted in a fairly high-pressure bypass flow. This flow was subsequently mixed with the flow through the high-pressure compressor with nine stages and the rest of the engine components to exit the engine through a

---

1. UK Patent no. 471368.

Fig. 57: A Pratt & Whitney JT3D turbofan engine nacelle with its separate fan flow nozzle on the Boeing 707 (military designation VC-137) that flew as Air Force One. Note also the inlet blow-in doors (that are variously positioned). (Photo by author at the Museum of Flight).

common nozzle. Historical photos of this engine usually show it with a rapid mixing nozzle for noise emission reduction (see Fig. 61, page 154).

The Rolls-Royce Conway set the stage for wide adoption of the turbofan. In 1961, Pratt & Whitney quickly introduced the JT3D, a turbofan engine of their own based on the JT3C turbojet for the 707 and the DC-8, both of which were gaining wide acceptance from the air travelling public. The JT3D had a two-stage fan that ran with a six-stage low-pressure compressor and further compression by a high-pressure compressor. That design was such a great improvement that replacements of the JT3C with the JT3D turbofans on 707s and DC-8s were carried out as fast as Pratt & Whitney could supply the engines.

For American Airlines, for example, the performance improvement from the JT3C-7 to the JT3D-1 was dramatic: take-off thrust increase from 12,500 to 17,000 lbs and about a 13 per cent improvement in fuel consumption.

The availability of new engines also allowed Convair to try to salvage their situation with the design of a new aeroplane. A larger aeroplane, the Model 990,

Fig. 58: Swissair Convair 990 in flight. Note the 'shock bodies' on the upper surface of the wings. The lower image is an inlet view of a GE CJ-805-23 nacelle showing the outer annulus for the bypass air and the central engine core inlet (San Diego Air & Space Museum).

based on the 880 and named the 'Coronado' (149 passengers), was introduced with a purchase commitment from American Airlines. Like the 880, it was designed to accommodate 2+3 seating. The CV-990 included the novel use of 'anti-shock' bodies on the upper surface of the wings. These bodies were intended to raise the critical Mach number[2] of the aeroplane and allow it to operate at Mach numbers around 0.9 as the 880 supposedly did. In practice it was often flown at lower speeds to increase the aeroplane range by reducing fuel consumption rate. The 990 joined the increasing trend to use turbofans with GE CJ805-23 engines.

The General Electric CJ805-23 engine was a significant detour on the road to today's commercial fan jet engines. In the late 1950s, the time when Pratt & Whitney was offering JT3D turbofans, GE stepped into the competition and thought they might advance the state of the art by building a turbofan around the military J79. The principal innovation centred on the location of the fan on the engine. The Pratt & Whitney, as well as the Rolls-Royce fans, were located on the front of the engine requiring a long shaft to convey the power to it from the turbine at the rear. GE's idea was to build the fan directly on the outside perimeter of a low-pressure turbine, obviating the need for any shaft. This engine modification consisted of attachment of a module to the standard single-shaft CJ805 engine. The module was a low-pressure turbine with an *integral* outer fan. By integral is meant that a blade consisted of a turbine 'bucket' close to its root and a fan blade further out. GE engineers called the combination a 'blucket'! The fan was also a single transonic stage that was lighter than the two-stage fan on the JT3D. The CJ805-23 engine featured a bypass ratio[3] of three, which was quite large for the time. The -23 engine was simple in concept and light in weight.

The JT3D, the Conway, and the CJ805-23 had similar thrust ratings of 16–17,000 lbs.

Before his ouster, Howard Hughes had wanted this advanced GE engine for the Convair 990s in the TWA fleet and so limited access to the first production group of engines for other airline operators. This was another attempt at controlling the engine market. Fortunately for buyers seeking the turbofans, units from Pratt & Whitney were available, and unfortunately for GE they turned to Pratt & Whitney.

The CV-990 was ordered by American Airlines to serve the Los Angeles to New York route. In practice that was not possible because the 990's range fell short. As a result, American reduced its order and sales momentum was lost. The production of the 990 ended a year later (1963), and with it a major customer

---

2. The Mach number where wave drag effects become significant.
3. Bypass ratio is the mass flow rate through the fan divided by that through the compressor.

Fig. 59: A GE CJ805-23 aft fan turbofan engine. The fan was designed as a module to be attached to a standard CJ805. Note the three engine mount points. The attachments for mechanical power devices (hydraulic and electrical) are visible under the inlet. The figure at right is of the rear showing the fan flow through the blue walled duct and the primary hot flow through the red centre. The lower figure shows a single fan/turbine blade as mounted in its rotor attachment. (Colour photos by Rick Kennedy, all photos courtesy General Electric, Evendale, Ohio).

of the GE aft-mounted-fan engine. A total of thirty-seven CV-990 aeroplanes were produced. While the sales of the Convair aeroplanes were insufficient for the company's economic success, they were indeed sold to a variety of operators. The market sector for which the CV-990 was designed was soon dominated by the Boeing 720 (with JT3Cs) and the 720B (with JT3Ds).

To contrast the Convair production numbers with those of the other four-engine airliners, we note that Boeing produced 865 model 707s in various versions (1958–79) and 154 of the 720s (1960–67) while Douglas built 556 DC-8s (1959–72).

The GE publication *Eight Decades of Progress* cited in the bibliography tells the Convair entanglement matter from their viewpoint:

The Convair 880 had not entered commercial service until May 1960, and sales of the airliner were not even close to the original market projection of 257 aircraft over the life of the programme. As *Fortune* magazine put it in a lengthy two-part story about General Dynamics and the 880/990 program in January and February 1962, General Dynamics/Convair had become 'entangled' with the 'capricious Howard Hughes.' The magazine quoted one GD executive: 'The 880 was an advanced plane with a better engine than any other at the time [1956–59]. Hughes wanted to keep it from TWA's competitors. So, people who might have bought the 880 if we [Convair] had been allowed to sell it to them, bought the DC-8 or 707 instead.'

According to *Fortune*, when Hughes ordered the first 30 of the new Convair jet transport, there had been 'an understanding [between Hughes and Convair] with Howard Hughes [that] had kept Convair from selling the 880 to anyone but TWA and Delta for a whole year.' Because of the 'understanding' and the loss of the United order, the market potential had dropped to 80 airplane.

The financial losses incurred by General Dynamics in connection with the 880/990 programme were thought to be the largest to date by any corporate undertaking.

Caravelle

The end of the Convair saga caused GE to attempt another road to enter into the commercial airliner market. In 1961 it partnered with French airframe manufacturer Sud Aviation to offer a CJ805-23-powered twinjet. Sud's Caravelle airframe was a small aeroplane with eighty seats. The Caravelle enjoyed commercial success since beginning service in 1959. Rolls-Royce Avon turbojet engines were the standard engine offering during the early years and Pratt & Whitney JT8Ds were installed on later models. The unsuccessful GE/Sud partnership was terminated after negotiations for a sale of twenty-five aeroplanes to TWA came to naught. The environment in which these negotiations failed included American Airlines and Eastern Airlines placing orders to launch the new Boeing 727. The last CJ805 engine was shipped in 1962. The last of the 282 Caravelles built rolled off the assembly line in the early 1970s as Sud (later Aerospatiale) shifted its design and manufacturing focus onto the supersonic Concorde with its partner British Aircraft Corp (later British Aerospace).

This tale of Convair as a major airliner builder closed with these events. Left were Boeing, Douglas and Lockheed. The latter had not entered the field in an effective manner, although it would again, for a time. Of the former two, The Boeing Company was most successful because it had the resources to offer aeroplanes of varying capabilities that the airlines wanted. Douglas, on the other hand, had to hope that airlines wanted what Douglas offered with its DC-8. That circumstance would determine the evolution of the airliners offered by these American companies in the years to come. For the time being however, the engines being developed had a strong influence on the shape and function of the airliners.

The introduction of the DC-8 by Douglas coincided with the retirement of Donald Douglas, Sr. from the company (1957). His son, Donald, Jr., who had been vice-president, rose to become chief executive and served in that capacity until the merger with the McDonnell Corporation in 1967. Donald Jr. did not share his father's interest and acumen in the aviation business, but was active in other fields, banking and yachting among them. A number of senior management and engineering people left Douglas in that period and, according to those who were there then, it was 'the beginning of the end' for Douglas. To the company's credit were, however, the successful design and sales of the DC-9. The DC-9 was a rear-engined twin with T-tail. It was medium sized with 2+3 seating and had a medium range designed to complement the larger DC-8 offering. It first flew in 1965 and entered service shortly thereafter. The last of the almost 1,000 DC-9s built rolled off the assembly line in 1982. The aeroplane's design went on

Fig. 60: A Soviet Aeroflot Il-62 airliner, with low bypass turbofans. (Peter M. Bowers Collection/Museum of Flight).

to become the basis of other aircraft produced by the companies that merged with Douglas.

A number of entries into the airliner market were made from outside the US in that time period, notably by Great Britain and the Soviet Union. The British Vickers VC-10 differed from the rather similar designs of the 707 and DC-8 in that the Conway turbofan engines were located on the sides of the rear of the fuselage in two pairs under a large T-tail, similar to that of the DC-9. This design decision might have been influenced in part by the need to consider patents associated with the American designs. The fifty-four VC-10 aircraft produced between 1962 and 1970 were used primarily for British Empire routes from the homeland. In the Soviet Union, the Ilyushin IL-62, a similar design to the VC-10, was introduced into service in 1967 and produced until 1995. It was primarily used by the Soviet bloc client airlines. Fewer than 300 of these aircraft were built.

Why is the Turbofan so much better?

The better fuel consumption of these fan engines was not their only advantage. There were two more and they accelerated the use of jet airliners into the travel business in very important ways. One was the reduced noise associated with the

Fig. 61: A daisy petal mixer nozzle on Boeing 707 powered by Rolls-Royce Conway engines. Note the thrust reverser cascade just ahead of the nozzle. (Boeing Images BI411462).

fan engines. The turbo*jet* required complex and costly means to reduce noise as much as possible with nozzle designs that promoted rapid mixing of exhaust and freestream air. These included so-called 'daisy-petal' mixers[4] and *multi-tube* nozzles that robbed the engine of some of its thrust. The turbofan was still noisy but significantly less so.

The third advantage associated with the use of turbofans is that a fan engine designed to produce the same thrust as a turbojet engine in cruise, automatically has more thrust at take-off conditions. For example, the Boeing 707-120 was introduced with four *turbojet* engines that each produced of 13,000 lbs of thrust while the later and identical 707-120B had JT3D *turbofan* engines producing 18,000 lbs of thrust at take-off. This was an impressive improvement. Besides accelerating the aeroplane off the runway more quickly, turbofan performance permits use of a significantly shorter runway, a boon to airport operators. Further, the consequences for flight operations are that high-altitude airports can be served more readily with fuller loads and with fewer restrictions associated with operation on hot days.

The turbofan was here to stay.

Bypass

The technical idea of the bypass air is to decrease the kinetic energy in the jet because the power associated with jet speed higher than flight speed is a waste. If it were not for that waste, propulsion efficiency would be perfect. In other words, all the power generated by the engine would be transmitted to the aeroplane. Just as the excess energy in the propeller wake of a ship is lost, so is that in real aircraft propeller streams. For typical aircraft propellers, that efficiency is between eighty and ninety percent because the jet stream is very large. The propulsion efficiency of a turbojet (without bypass) is much lower, on the order of fifty percent. The turbofan, on the other hand, raises this number towards significantly higher values. That is the underlying principle at play and the motivation for the developments about to be described.

The first production turbofan engine, the Rolls-Royce Conway (RB.80), was designed and developed in the mid-1950s and had a relatively small 'fan' airflow rate as mentioned above. It processed about twenty-five per cent more air through the so-called fan (the LP compressor) as went into the HP compressor and the

---

4. 'Noise Suppression Nozzle' Patent US3053340A by John T. Kutney (General Electric Company). Applied for 1958, granted 11 September 1962.

following components. In the industry, one speaks of this fraction as the *bypass ratio* (0.25 in this case). Various sources quote bypass ratio values for the early versions of this engine as high as 0.6. Whatever the value, it was a rather timid start for improving the engine's performance, a start nevertheless. As the Conway engine evolved, the bypass ratio was increased and the value of the initial bypass ratio will vanish in significance.

In practice, the Conway installation on a 707 yielded a significant fuel mileage improvement for the aeroplane. For American airlines, the option to buy British Rolls-Royce engines was, however, not viewed as a politically practical option.

The switch from JT3Cs to the turbofans was, for Pratt & Whitney, tough. It came at a time when the investment in turbojets had not even begun to be paid back. It took competitive pressure from GE with their CJ805-23 and threats from the airlines to switch to the Convair 990 for Pratt & Whitney to realize the seriousness of the situation. The airlines were operating 707s and DC-8s for which the option to use American turbofans was not (yet) available. For Pratt & Whitney management, part of the challenge of the turbofans was that the market for turbofans was primarily driven by commercial and not by military engine needs. The normal way of things had been to adapt military engines to civilian aeroplanes, a process that greatly reduced the development funds the company had to lay out. Now the monies invested had to be justified by the new world of jet airliners that, at that time, was far from settled. That evolution of events would be repeated with Pratt & Whitney's JT9D engine during the development of the Boeing 747 in the period between 1966 and 1969. There would be no customer for the engine except the 747.

Pratt & Whitney management recognized that the significantly superior performance of turbofans could not be ignored, and delaying or not acting on the information was risky. The company took rapid steps to configure the turbojet into a fan and the situation was saved when viewed from a historical perspective. The first run of the Pratt & Whitney JT3D was in 1958 and it entered service in March 1961 on the Boeing 707-120B. The bypass ratio for that engine was a more ambitious 1.42.

The conversion of the JT3C turbojet to the JT3D turbofan involved changing the low-pressure portion of the engine that was a 9 LPC driven by 2 stage LPT to a 2 stage fan and 7 stage LPC driven by a 3 stage LPT. The high-pressure section was unchanged. Figure 57 (page 148) shows the JT3D configuration with its separate bypass flow outlet. The choice of a short cowl for the JT3D, for example, was made to allow for a quick redesign by keeping all the accessories mounted on the engine where they were near the waist of the engine just ahead of the burner

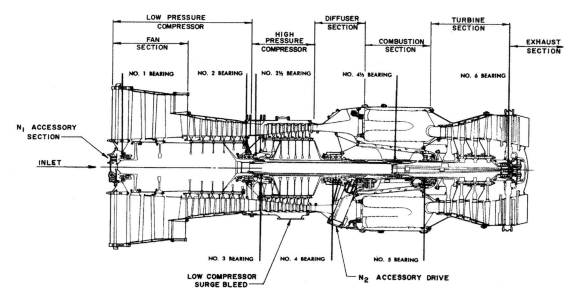

Fig. 62: Cross-section view of a P&W JT3D turbofan showing details of the various engine elements. Also shown are the locations of bearings supporting the two internal shafts. Note the accessory attachment points and the surge bleed valves on the high-pressure compressor section to prevent surge in the low-pressure section. (Credit: K.F. Whatley, ASME paper 62-GTP-16)

(see Fig. 53, page 132). The JT3Cs in service were, in some instances, accepted by Pratt & Whitney as 'trade-ins' for the JT3Ds. In addition, Pratt & Whitney offered kits for airlines to upgrade the engines into turbofans. In that same time period, GE was developing its aft fan engine with an even more ambitious bypass ratio of 3.

Improving performance meant exploiting bypass technology further. In Britain Rolls-Royce built a successor to the Conway, the Spey (RB.163), with a larger, but still modest, bypass ratio of about 1.0. This engine was also produced in the US as the Allison TF41 and, as such, was justifiably called an American engine. It powered military aircraft as well as the British Aircraft Corporation BAC One-Eleven, a short-range twin-jet airliner produced in two versions between 1965 and 1989 with seating capacities of 89 and 119.

The race was on to improve the performance of commercial jet engines. New aeroplanes were being developed to replace older fleets of smaller aircraft and they needed new engines. In the late 1950s, Boeing, for example, was working on the 727 and British engines were considered to power it. Pratt & Whitney felt the pressure to widen its engine offerings beyond the JT3D. They turned to a one-trip (expendable) turbojet engine, the J52 (company name: JT8A), designed for the Hound Dog cruise missile. In that application, the resulting engine had a movable centre body nozzle not unlike the Jumo 004 engine. In view of other military development needs and the need to use private funds, the decision to embark this new engine development was again not easy for Pratt & Whitney. Such steps are always costly and risky. In 1960 the decision to go ahead was made, and it became a new commercial turbofan engine to be named the JT8D. Pratt & Whitney company specifications stated that the JT3D consumed 0.78 lbs of fuel per lb of thrust per hour at cruise conditions (35,000 ft altitude, 4,000 lbs thrust level, and Mach 0.82 flight speed). The new JT8D was developed to deliver a claimed 19 percent better fuel economy. The JT8D engine was somewhat smaller than the JT3D as measured by intake airflow rates estimated to be 330 lbs of air per second, compared to the 3D's 500 lbs per second at sea-level, static conditions.

The JT8D was developed in two phases with early model numbers 1 through 17 producing between 12,000 and 16,000 lbs of thrust. It had a rather modest bypass ratio and a strong fan pressure ratio. It was introduced into service with the Boeing 727 in 1964. Unlike the JT3D with a separate exit flow nozzle for the fan (see Fig. 57, page 148), the JT8D had a common exit nozzle for mixed core and fan flows (Fig. 64). This was a requirement imposed by the installation in the centre engine position in the 727. Happily, this also made for some reduction in the noise emitted from the engine as a whole. Note in Fig. 63 that the accessories on the JT8D were outside the fan case to allow the bypass flow to hug the core without obstructions.

Starting in the 1970s, changing environmental regulations led Pratt & Whitney to reconfigure the JT8D into a -200 series to make the engine more efficient and

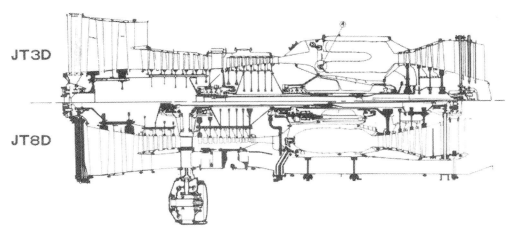

Fig. 63: Schematic comparison of the two P&W turbofans. Note the more compact combustor in the later JT8D engine configured before the -200 series.

less noisy. It is safe to say that component performance had been improved in the time between the 3D and early 8D engines, so that the JT8D-200 could be designed with improved performance and increased thrust of over 21,000 lbs. In the -200 series most of the improvements were made outside the well performing high-pressure core components of the earlier model. Figure 64 shows the cross-section of the two JT8D versions in contrast to each other with the engine core common to both. Interestingly, the collection of jet engine images shows the progress made in time in reducing the volume necessary for the combustion chambers. The combustors were becoming an ever-smaller, less-noticeable part of the entire engine.

A widely used JT8D-219 model was introduced into service in 1980. For the efficiency improvement, the bypass ratio was increased to 1.74 (from .96). For the noise, the fan pressure ratio was reduced from 2.21 to 1.92. We note parenthetically

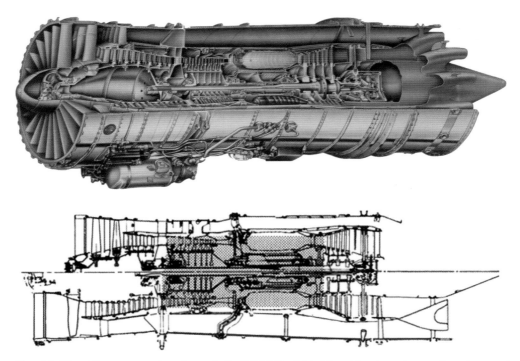

Fig. 64: Top: Cutaway engine view of a P&W JT8D-200. Note blue long bypass flow duct and the two-stage fan ahead of it. A mixer blends the fan and (red) core flows at the exit. (Ronald Lindlauf, New England Air Museum). Bottom: Schematic engine view of the conversion of the early model of the P&W JT8D (upper image) to the later JT8D-200 (lower). A single stage fan replaces a two-stage fan with the shaded high-pressure core common to both models. The mixer on the -200 model combines the core and fan flows quite thoroughly. (Pratt & Whitney Archives).

that a pressure ratio of 1.89 leads to sonic flow and a noise crackle associated with shock waves emanating from the boundary between the jet and the freestream air. At the higher fan pressure ratio, the jet noise was a significant component of the total noise emitted at take-off conditions. That noise is hardly ever to be heard now that low bypass ratio engines are rarely used in commercial service.

The decision to pursue the JT8D development turned out to be a very good one for Pratt & Whitney. The engine lived a long and productive life as power plants for the 727, the 737, and the DC-9 and its successors and was a major commercial success. The company produced more than 14,000 of the early engines and about 2,900 of the upgraded -200 version by the time production ended in 1996.

With the risky and financially demanding step of developing the JT8D, the company became the premier American, if not global, producer of jet engines for commercial aircraft.

Full Reverse!

With the Conway engine, Rolls-Royce pioneered another aspect of engine design adopted universally in all engines today. They recognized the usefulness of a propeller-driven aeroplane's ability to change the pitch on the propeller to obtain reverse thrust or at least adjust it to produce a lot of drag after touchdown. It shortens the landing roll and saves wear and tear on brakes. With this idea in mind, the RB.80 Conway engine incorporated a thrust reverser. The reverser involves blocking the rearward flow of the jet(s) through the propulsion nozzle and partially redirects it forward to obtain about forty per cent of the forward thrust in reverse. These reversers were deployed either with an external redirection using

Fig. 65: Left: Clam shell or target thrust reverser on an MD-80 aeroplane with P&W JT8D-200 series engines. (wikipedia). Right: Cascade reverser on a 727-100. (Photo by author at the Museum of Flight).

The Turbofan Engine   161

Fig. 66: High bypass fan powered airliner (Boeing's prototype 747 on display at the Museum of Flight) with no. 3 engine configured to show thrust reversing. The lower picture shows the flow blocking 'doors' in the fan case to allow the reversed flow to exit via the green cascade in the upper picture. The core flow is not reversed. (Photos by author at the Museum of Flight).

what is called a target vane (Fig. 65) or an internal cascade that may be exposed by movement of the engine cowling to the rear and simultaneous blocking of the rearward flow to the propulsion nozzle (Fig. 66). In modern high-bypass engines, use of the cascade reverser is almost universal, though not necessarily on all engines of four-engine aircraft, and used only on the fan-flow since the core engine flow contributes relatively little to the total forward thrust.

Jet Helicopters

The advent and virtues of turboshaft engines were not lost on the helicopter industry, that saw that the power per unit engine weight was so much better for the gas turbine engines than for the reciprocating engines of the post-war period. In the Sikorsky S-55 and S-58 helicopters described earlier, the engines were mounted low in the front of the helicopter with a relatively long shaft that conveyed the power to the rotor transmission. That made for good engine maintenance access and for keeping the aircraft centre of gravity low. The light-weight turboshaft engines could be mounted high and near the rotor transmission. An additional advantage was that its maintenance requirements were modest compared to that of the reciprocating engines and the configuration did not severely compromise the practicality of servicing the engine.

That high location also allowed thinking about designs for helicopters that could 'land' on water, a feature that had some attraction for Sikorsky with his experience with flying boats. He designed two amphibious helicopters that were very successfully deployed in the US Navy and US Coast Guard. These were known by the company names of S-61 and S-62 and developed in the early 1960s. The S-62 was a single engine version of the S-61. As for the fixed wing aeroplane, the gas turbine employed in a helicopter made for a beautiful marriage between engine and aircraft.

The S-61 and S-62 had a number of different labels as US military aircraft and evolved over the years. They were also sold to civilian and military customers around the world. The basic design of the follow-ons remains that of the S-61. As Marine One, such aircraft still serve to convey the President of the United States from the White House to various nearby locations, among these, the airport where Air Force One may be parked.

In the jet age, Sikorsky built a small number of helicopters based on the S-61, large enough to be 'airliners'. These were aimed at the limited markets available, such as from the downtown area of large cities to their outlying airports. The economics of operating helicopters as airliners are, in practice, difficult. Consequently they

The Turbofan Engine   163

Fig. 67: A Sikorsky S-61 airliner powered by two 1500 HP turboshaft engines located above and to the rear of the cockpit. Date is 1960. Capacity is for 28 passengers and cargo. (Igor I. Sikorsky Historical Archives ©2018).

are not frequently seen as a practical or even available mode of public transport. To this day they do, however, remain as a component of the air travel industry with newer and market tailored aircraft.

The engines for the Sikorsky helicopters were almost exclusively GE T58 turboshaft. This engine first ran on a navy version of the S-61 (HSS-1) in 1957 and produced about 1200 HP. It featured a purely axial flow compressor and a straight-through annular combustor. These features make for a long and slender engine and an interesting contrast to the Lycoming T53 shown in Fig. 71 (page 174). The T58 was uprated over its long production lifetime that ended in 1984.

Bigger is better

The engine company AVCO-Lycoming produced turboshaft engines primarily for helicopters. It employed a number of the former Junkers engineers who designed the Jumo 004. It occurred to them to ask: since bypass is beneficial, would a greater amount of bypass be even better? The Lycoming turboshaft T55 engine already had a turbine that took out virtually all the available power from the engine gas flow as input to a gearbox. It was a rather simple matter of applying that power to a fan. After looking at the question analytically, the company built a demonstration engine using core components in hand matched to a new large fan. The promising analytical findings backed by experimental test stand results were reported in a professional meeting. The first run was in February 1964. A patent[5] for the idea was applied for in 1965 and granted in 1968.

The high-bypass turbofan looked promising.

Interestingly, an examination of the patent documentation reveals the use of a reverse-flow combustor engine, like the T53 (similar to the T55) shown in Fig. 71 (page 174), to illustrate the concept as it was materially demonstrated. In nearly all subsequent designs of high-bypass engines, the combustor arrangement would be more conventional.

Big Bird – Big Propulsion

In the 1960s, turbofan technology was developing to allow the US military to consider building a really large transport aircraft to carry bulky military cargo to remote places around the world. To that end, proposals from all the major airframe and engine manufacturers were solicited and received in 1964. These proposals were narrowed to airframers Boeing, Douglas, and Lockheed and to the engine-makers Pratt & Whitney and GE with study contracts to refine the design of an aeroplane able to carry the projected loads of over 100 tons and meet a variety of performance requirements. The new aeroplane was going to need powerful new engines. Boeing, Lockheed, and Douglas submitted proposals to build this aeroplane and Lockheed (that later became Lockheed-Martin) won the competition with a design to be called the C-5A. Both the Boeing and Lockheed designs had the cockpit on a higher level so that payload bay access was by means of a door on the aeroplane's nose. The resulting C-5 aeroplane was produced in two batches: eighty-one C-5As built in the years 1968–73 and fifty C-5Bs between 1985 and 1989. The C-5 entered service in 1970 and still serves today.

It is the engine side of that story that is even more interesting. General Electric developed the first production high-bypass turbofan in connection with the C-5. The engine was the TF39, a true pioneer in the development of large engines. Its

---

5. The technical paper is ASME paper 64-GTP-15 'Potential of the High By-Pass Fan', by S.H. Decher and D. Rauch, March 1964, and patent no. 3,390,527 'High Bypass Ratio Turbofan' granted to the same persons on 2 July 1968.

Fig. 68: A display model of the GE TF39 turbofan showing the exits of the two fan flow streams. These flows exit via a common nozzle. (Photo by author at General Electric in Evendale, Ohio).

development was roughly in the same time period as the work of the Lycoming engineers and might have resulted in GE successfully patenting the high bypass concept.

The TF39 was powerful, with 41–43,000 lbs of thrust, and innovative. It had an unheard-of bypass ratio of eight with a unique 1½ stage fan: the outer half of the inlet air flow is compressed by a single-stage with inlet guide vanes, while the inner half of the fan has two stages without inlet guide vanes. The engine design used the von Neumann variable stator control (capitalizing on the experience with the J79 and CJ805 engines), advanced turbine cooling, a cascade thrust reverser on the fan, and mid-span vibration stabilization on the large fan blades. These so-called 'snubbers' prevent blade flutter and fatigue failure. The engine first ran in 1964 and set the stage for future large turbofan engines.

While Lockheed and General Electric were pioneering the military C-5, Boeing took its design and turned it into a civilian airliner to become the 747. Boeing went on to select the engine that Pratt & Whitney had proposed for the C-5. That engine became the JT9D that later powered a number of other airliners, including the Airbus A300 and A310, the Douglas DC-10, and the Boeing 767. It seems losing a military competition is not necessarily a bad outcome … if you have the resources to take advantage of the engineering work done for the military contract.

By 2017, Boeing had built over 1,500 747s. By the end of the production run in 1990, Pratt & Whitney had built more than 3,200 JT9D engines. Subsequent 747s were powered by GE and Rolls-Royce, as well as PW4000 engines from Pratt & Whitney. The PW4000 family of engines, with thrust from 50 to 100,000 lbs, is a widely used derivative development of the JT9D.

GE's CJ805 experience included durability issues in service. These concerns caused the company to rethink its approach to engine design by focusing on a general-purpose engine core consisting of the combination of high-pressure compressor and turbine with the combustor in between. Such a core is often referred to as a 'gas generator'. The idea was to employ this core with the addition of low-pressure components and a fan to devise engines for a variety of capabilities. GE rightly credits Gerhard Neumann for this innovation and its realization. One application of this high-performance engine core was in the successful commercial CF6 high-bypass turbofan employed initially (1971) in the Douglas DC-10 and later elsewhere[6]. The CF6 became an effective competitor to the Pratt & Whitney JT9D. For the DC-10's competitor, the L-1011, Lockheed used Rolls-Royce RB.211s, engines of a similar design. That choice, by necessity as we shall see, helped to keep Rolls-Royce in the business with a committed customer for its engines.

The advantage of the high-bypass fan is recognized for its role in (again) reducing fuel consumption and noise and realizing the high thrust of the engine. The bypass ratio of the early engines of the high-bypass type was around five (with the exception of the GE TF39) and more modern engines are tending towards ten. All modern airliners use high-bypass engines today and that engine type certainly makes the large airliners practical in that four or fewer engines provide sufficient thrust. While aeroplanes with more engines have been built, it is hard to imagine them used as airliners. They would have been too complex and costly to operate.

Pan Am and the 747

As the high-bypass engines and wide-body aeroplanes became a reality, airlines with a look to their futures saw that the large passenger airliners were practical and

---

6. The list includes the A300, A310, A 330, Boeing 747 and 767, Lockheed C-5M, and the MD-11.

made sense from an economic viewpoint. Pan American Airways again stepped into the breach of progress and became the launch customer of the Boeing 747. Operations with the 747 started in 1970. Pan Am's story is, however, a sad one with its demise initiated by the 1973 oil crisis followed by unsuccessful attempts at financial recovery. The company ceased operations in 1991. Today Pan Am is no more and its history is not the only one to reflect the industry's difficulties that resulted in bankruptcies, mergers and other consequences. The Airline Deregulation Act of 1978 was an important ingredient in the turmoil that further affected airlines in the years that followed.

No story of airlines would be complete without mention of one of the visionaries in that industry: Juan Trippe. He founded Pan American World Airways in 1927 and in that connection wielded a great influence on the airline industry. He was a great admirer of the aeroplanes Boeing built and was the launch, or at least an important customer for, many of Boeing's models. These include the 314 Clipper, the 377 Stratocruiser, the 707, and the 747.

Pratt & Whitney was the first builder of high-bypass engines for airliners and the 747 was designed around their capability. New engines are, however, rarely without teething problems and the Pratt & Whitney JT9Ds had their share. Joe Sutter[7], chief engineer for the 747 described the issues in his book. The central issue was compressor stall (or *surge* as it is often called), the ever-present issue in operating gas turbine engines. The nature of these stall event circumstances may never be available for historians but they did cause some engines to be classified as 'shaker' engines with deleterious consequences for the airframe and its operation.

There were also structural issues, fan shaft strength and fan case ovalization. Ovalization occurred under certain conditions, including on take-off rotation, because the fan case was initially insufficiently rigid. The undesirable consequence was contact between the fan blades and the case, not a good situation for safe aeroplane operation. As blade contact with the fan case is always to be minimized, the clearance between the two elements is also to be minimal to prevent leakage. The engineering solution to this challenge is to make the fan case inner surface of a material that can be safely scraped by the blades under extreme conditions. The clearance will, thereafter, be as small as can be manufactured.

These were new issues associated with the building of such a large engine and, to Pratt & Whitney's credit, they were successfully addressed.

---

7. Joe Sutter, *747: Creating the World's First Jumbo Jet and Other Adventures from Life in Aviation*, Smithsonian Books (Harper-Collins), 2006.

For Boeing, the engine issue also involved a mid-design increase in the required thrust of several thousand pounds because the aeroplane gross weight had grown from 625,000 to 710,000 lbs as it was being developed. In the end, the improved 47,600 lb-thrust JT9D-7A engines served well, despite the rough economic times in the early 1970s when the 747 began service and became the flagship for many an airline.

The history of the cooperation between Boeing and Pratt & Whitney during the development of the 747 is described differently by Joe Sutter and by Jack Connors writing about their sides of the same story. In Connors's Pratt & Whitney version, the teething problems of the JT9D are hardly mentioned, while they loom large in Boeing's view of the time. For Boeing, the consequences of fixing engine problems involved delays in the delivery of the engines while the airframes were being built and stored on the field without engines. Boeing was not getting paid until they delivered a whole aeroplane, making for a stressful cash-flow situation.

Trouble: the Rolls-Royce RB.211

As Pratt & Whitney succeeded building 'dependable' JT9D high-bypass turbofan engines for airline service, GE with a similar CF-6 engine was close behind. The market for these engines was competitive. Was there room for a third competitor? Rolls-Royce wanted to be in the game and developed the RB.211. This turbofan was the logical development of the Spey. The company introduced the RB.211 in the early 1970s where it was used as the engine for the Lockheed L-1011, the Tristar, a wide-body aeroplane not unlike the DC-10. In various versions, this novel, three-spool engine continued as engine of choice for many other commercial aeroplanes until the late 1980s. That journey was not without a serious bump in the road.

In an attempt to surpass the performance of the American engines, Rolls-Royce undertook a development programme to use a lighter carbon fibre fan instead of the solid titanium alloy fan used by the Americans. Unfortunately the carbon fibre fan is vulnerable to damage at the leading edge from sand, dust, and material of that kind. Rolls-Royce went on to try to solve that problem by attaching a metal leading edge. That challenge overwhelmed Rolls-Royce. The edge was difficult to keep in place and continued efforts to resolve the problems were ultimately unsuccessful. So costly were they that Rolls-Royce was forced to declare financial insolvency. The British government was not anxious to lose the industrial capability that Rolls-Royce represented and in 1971 purchased a majority interest in the company, so it became Rolls-Royce, plc. Considering the events that followed, the decision to financially support Rolls-Royce in time of need seems to have been a good one. Today Rolls-Royce is again in private hands and is the second largest producer of large turbofan engines after General Electric.

The first generation of high-bypass turbofans, the Pratt & Whitney JT9D, the GE CF-6 and the Rolls-Royce RB.211 (all with solid metal fans) powered the large jets that followed the introduction of the Boeing 747, the Douglas DC-10, the Lockheed L-1011, and others. These bypass ratio five engines had thrust levels between 40,000 and 60,000 lbs. They also had fan pressure ratios around 1.5 that, at take-off, kept them well below the threshold associated with sonic flow noise from the fan nozzle.

One aspect of the design of a gas turbine engine, including the turbofans, is the engine designer's choice of pressure ratio in the compressor. In modern turbofan engines this value can range from the mid-20s to over 40 by providing the appropriate number of compression stages. 'Low' values of the compression ratio are used in engines that are intended to be light in weight and powerful. 'High' values are required for efficient (good fuel mileage) operation. A specific airline's needs may be for engines to service long distance routes where fuel economy is important. On short hops, a light, powerful engine may be better. Hence an airline might choose from among one or another manufacturer's offerings to be able to employ aircraft-engine combinations tailored for use on specific routes. In practice, intermediate values may be a good compromise to allow for some dispatch flexibility.

Smaller Big Fans

While the 'big three' produced the large high-thrust turbofans, high-bypass engines were also adopted for smaller aircraft. Notable among these is CFM International, a partnership between GE and Safran (named SNECMA at the time). The consortium produced CFM56 engines (thrust levels 20–23,000 lbs) for the various models of the Boeing 737, among others. Another of the engine manufacturing consortia is International Aero Engines, created in 1983. IAE has had a temporally fluid membership from the US (Pratt & Whitney), Great Britain (Rolls-Royce), Japan (Mitsubishi), Germany (MTU), and Italy (Avio). It produces the V2500, a 25,000 lb-thrust turbofan for the narrow-bodied Airbus A320 series and the MD-90[8].

---

8. The MD-90 is a later version of the Douglas DC-9 produced after Douglas merged with McDonnell Corp. in 1967. The McDonnell-Douglas Corporation subsequently merged with Boeing in 1997. This aeroplane was marketed for a time thereafter as the Boeing 717.

Even Bigger!

Since high-bypass was such a good idea, higher values are even better when the basic engine can perform well enough to drive a bigger fan. Indeed, the development in the 1990s allowed two manufacturers to develop a new class of engines with bypass ratios near ten: the GE90 and the Rolls-Royce Trent family of engines. An earlier Trent (RB.203, a low bypass engine) is not part of that family. This newest Trent engine is built in many variants, with thrust levels ranging from 65,000 to almost 100,000 lbs and bypass ratios ranging from five to ten. Both the Trent and the GE90 are competing for application in the large-engine market. The latter can be obtained with a rating of 115,000 lbs of thrust!

The Fan

A larger fan presents design challenges: the fans are about ten feet in diameter. Rotation requires that the outer parts of the blades travel faster than the inner parts but an important goal of the design is to achieve an outlet pressure that is as uniformly high as possible. That aspect of propeller-, compressor blading-, and fan-design has always been an issue for aircraft-propulsion machinery. The propeller blades on piston engine aircraft were made twisted and variously cambered (curved in blade cross-section profile) so that this goal is realized. Similar measures are incorporated into fan blade design with one important difference: the fan has an inlet in front of it.

The inlet of a jet engine slows the incoming air down a bit, so that the 'flight' speed component of the air impacting the fan blade is reduced somewhat and sonic flow effects are consequently reduced. Ultimately however it is the high speed of the outer part of the fan blade motion that sets limits on what the fan can do efficiently. In modern practice, the blade's outer flow field is supersonic (locally faster than the sound speed) and the inner portion is subsonic, with a transonic region in between. The aerodynamic behaviour of these flow regions differs significantly in that the waves mentioned earlier in our tale play roles that vary: relatively unimportant (near the rotor hub) to extremely important (at the tip). Furthermore, the mathematics required to describe these flow regions differs substantially and, to make things really difficult, the boundaries between the regions are variable depending on the flow conditions. Modern computers and the mathematical description of the flow behaviour that is programmed into them are up to the task and getting better.

Modern bypass fans, in contrast to compressors on turbojets, are not configured with inlet guide vanes, as the figures below illustrate. Rather, a fan exit stator is used to insure that the exit flow is only in the rearward direction and no kinetic jet energy is wasted in rotation.

Consider the blade shape of two high-bypass turbofan engines, one dating to the 1970s and the other to the 1990s. The figure below is of the front view of a GE CF6 on the left and a GE90 to its right. A blade of the GE90 is shown in isolation. This blade is a successful manifestation of the goal to build a lightweight fan blade that eluded Rolls-Royce in the 1960s. It has a metal leading edge for high tolerance for damage by material ingested by the inlet and a fibre composite remainder of the blade. If nothing else, the contrast between these two designs speaks to the advances made in understanding the flow field that the blades of turbofan engines deal with and the sophistication of manufacturing them.

Finally we can put all this discussion into a representation of an engine in its entirety. The figure below is of a GE90 showing the fan at the left, the narrow waist combustion zone further aft, and the wider turbine section at the rear. The relatively small size of the combustors speaks volumes of the technological progress that has been made when that size is contrasted to the early engines of Whittle and GE in the 1940s.

Fig. 69: High bypass fans from a CF6 (1970s) and a GE90 turbofan (1990s, to scale) with a single blade of the latter shown. Note the 'snubbers' near the tips of the CF6 fan blades. (Courtesy General Electric).

Fig. 70: A GE90-115B bypass ratio 9 engine. This engine operates with maximum rotational speeds N1 (fan and LP components) and N2 (high pressure components) at 2355 and 9332 RPM respectively. (Courtesy General Electric).

The picture partially illustrates how such engines are installed in aeroplanes. There are only three points where the engine is attached to the airframe. Two mounts near the front of the engine and a rear mount (clearly visible here) attach it to aeroplane structure. The attachment is by connections that break cleanly to allow the engine to separate from the aeroplane in a way that minimizes damage to the airframe if circumstances demand it. Fig. 59 (page 151) shows engine attachments on the CJ805-23.

Gearing – old and new

Is a bypass ratio of ten the limit of what technology can do? Possibly not. Pratt & Whitney, for example, is developing a family of engines, one of which, the PW1900G, sports a bypass ratio of twelve. Such a fan is sufficiently large in diameter, in relation to the basic engine size, that the rotational speed of the turbine shaft, which is dictated by rules of its own,[9] cannot be used in practice. The shaft

---

9. One of the rules deals with turbine tip speed limitations having to do with the centrifugal forces acting on the blade and its ability to withstand the associated stress.

speed for the fan must be lowered by a set of gears. Another member of the family, the PW1000G, is reported to have a gear ratio of three. Aside from turboshaft engines where gearing is necessary, the use of gearing has been avoided in the large engines. Gears consume power and pose potential reliability issues. The power handled by the gearing is not trivial because even a half per cent of mechanical energy loss in the gear box from a power flow of 30,000 HP is 150 HP of heat that has to be reliably removed from it. From an overall viewpoint however, such a heat loss is more than offset by the gain from the overall performance improvement.

Gearing has been used with success in the smaller engines powering regional jet airliners since the 1970s. Examples are versions of the Rolls-Royce-SNECMA M45, an engine designed specifically for the short haul airliner VFW-Fokker 614. In this engine, the notion of a variable pitch fan was also explored. Unfortunately its development was caught up in Rolls-Royce's financial difficulties in the early 1970s and was not carried further. Another geared engine is the Lycoming (now part of Textron, Inc) ALF 502 that was derived from the T55 turboshaft engine. This engine is used in the British Aerospace BAE-146 short haul airliner and a variety of business jets.

Large aircraft engines gave up the use of radial flow compressors in favour of axial flow compression. A couple of exceptions are worth noting in the small-engine sector of the industry. Lycoming's development of turboshaft engines was primarily for the US Army. Its helicopters often operate in dusty environments so that some design care had to be employed to avoid damage from ingested dust and sand. The reality of compressor design is that the last stages of blading in a purely axial flow compressor involve relatively small and thin stator and rotor blades. Ingested sand could cause unacceptable wear, particularly on the last stages. A Lycoming innovation was the use of a radial flow stage as the last element of the compressor. In such a design, compressor outflow is in the radial direction and that could compromise the engine configuration by making its diameter larger. Lycoming engineers took advantage of the situation however, to incorporate a reverse flow combustor allowing the turbine to be located much closer to the compressor. The result is a short, compact engine with a short shaft. The radial stage and the reverse flow combustor were used in the ALF502 engine as well as in the Garrett (now Honeywell) TFE731. These engines and the GE T700[10] engines found application in a large variety of business jets and helicopters.

---

10. The GE T700 engine does not use the reverse flow combustor and does employ (as do others) so-called 'blisk' axial flow compression stages where blades and disk are formed as one piece.

174 Powering the World's Airliners

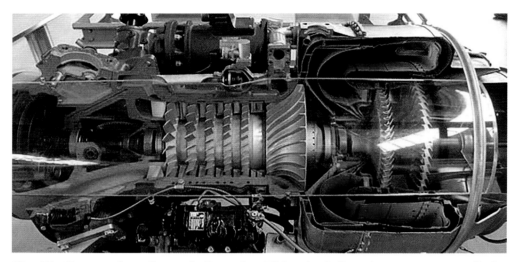

Fig. 71: Cross-section of an early Lycoming T53 turboshaft engine as a cut open display model. Airflow is from left to right. The T53 is a predecessor to the T55 of similar design. The T55 evolved into the ALF502 turbofan. Note the axial stages followed by a radial stage and the reverse flow combustor. The first (unshrouded) turbine is the gas generator turbine while the second (shrouded) is the free power turbine. The engine shaft rotates clockwise (viewed from the front) while the output shaft runs anticlockwise. Output gearing is on the far left. This engine design dates back to the mid-1950s. (Wikipedia: Sleipnir).

No Fan Duct?

Innovations are an inherent part of the evolution of aircraft engines. One interesting example from the 1980s is the GE36. It is termed an *unducted* fan because the external fan case is omitted altogether in the design. The combination of rotor and stator of a conventional fan is replaced by two counter-rotating fans that had to be capable of handling the relatively high Mach number external flow. A flight Mach of 0.75 was envisaged for the design. The two fans (as is the rotor-stator combination) are necessary to insure that axial inflow into the fan also leaves the fan axially to avoid any power investment in rotation of the flow. This engine included gearing by necessity. Pratt & Whitney and Allison partnered to produce a competitive design. Neither so-called 'propfan' found a military or commercial application.

The future will likely see further improvements in engine performance. While predictions are risky, the use of new materials such as aluminum and/or fibre composites to lighten the fan further is quite realistic. As the performance of a gas turbine depends on relatively few parameters, efforts at improving them will invariably involve better understanding of the flow through the components to minimize losses and on increasing the turbine inlet temperature to higher levels yet, levels that already exceed the turbine blade material's ability to withstand the heat without active and very sophisticated cooling. It has to be sophisticated because the cooling air is an expensive part of the power budget of the engine.

Wings are like Jet Engines!

Jet engines have evolved to move more and more air, pushing that air ever more gently to avoid investing a lot of excess energy in jet exhaust. That is the goal of the bypass engines. The same thinking also applies to the wings of the aeroplane. Consider that the purpose of the wing is to push a mass of air downwards to generate lift. This can be done with a short wing pushing a modest amount of air but pushing it hard. Alternatively, the wing could be made longer in span with the result that more air is affected by the lifting process but the energy invested in the downward moving air is smaller.

The momentum and the energy inherent in the downward flow (called downwash) are determined by the air speed, or more correctly velocity, as the wing pushes air downwards. Specifically, the momentum changes that result in lift vary directly with the downwash velocity. The energy, however, is much more strongly determined by that same velocity *squared*. The difference in speed dependence by momentum and energy is at the heart of (mathematically) stating the efficiency of providing lift and, as discussed above, thrust from the jet. In short, what is wanted is momentum or momentum changes to realize forces like thrust or lift. That result has to be paid for with an expenditure of energy. For the aeroplane as a system, its efficiency as measured in fuel mileage is strongly influenced by the manner in which lift and thrust are realized.

In cruise, the downwash from a wing varies directly with flight speed as the wing deflects the freestream airflow by a small angle. By contrast, it is relatively large when the aeroplane flies slowly to take off or land carrying the full weight of the aeroplane. The wing design allows that by having the ability to change the wing's camber, *i.e.*, its shape curvature. The wing configuration changes the flow direction by a few degrees in cruise, and by about 45 degrees when travelling slowly to maintain lift.

The parameter in aeroplane aerodynamic design that describes the wing's induced drag, *i.e.*, resistance due to lift associated with the downward flow from the wing, is called the *wing aspect ratio*, or *AR*. It is roughly[11] the ratio of span to

---

11. Technically, *AR* is the ratio of squared span to wing area. This formulation accounts for wing taper and other aspects of its shape.

Fig. 72: Top: Leading edge slats and trailing edge flaps in a landing configuration of the first Boeing 727 (-022). Note that air is deflected by about 45 degrees. Lower photo is of the extended Kruger flap on the inboard side of the wing leading edge. (Photos by author at The Museum of Flight).

chord where the chord is the length of the path travelled by the air over the wing. A glider has a very high AR, in the range of 25–50, while a jet fighter might have an AR of 2–3. Extremely high AR, while good for minimum lift-induced drag, is not desirable in a jet airliner because of the surface friction that accompanies the extra wing surface area. Thus the design challenge for the aerodynamicist is to balance lift-induced drag with drag associated with skin friction on the aeroplane's various surfaces to minimize the total. The choice of AR is the main tool to do that.

To put this parameter into some perspective, we note that the Boeing B-29 and the Model 377 Stratocruiser's wing aspect ratio was about 11.5, not a glider but designed for long range. The higher the aspect ratio, the better is the all-important performance parameter, lift/drag, or L/D. That measurement (actually its inverse) states how much drag has to be dealt with to provide a pound of lift.

The first generation of long-range airliners, the 707 and the DC-8, were built with aspect ratios near 7 and realized an L/D of about 18, meaning that the drag force was about 5 per cent of the aeroplane weight. The 747 wing aspect ratio increased from 7, to 7.9, and then to 8.5 as its configuration evolved. At the present time (around 2018), the 787-8 has an AR of 9.59. The achievement of an L/D over 20 is the likely result. The latest version of the proposed 777-9 exceeds that with AR of 11, very much like that of the B-29 but flying much faster! In that instance, the wingspan is so large that it is proposed to have the outer part of the wings fold back (like naval carrier-based fighter aircraft) to be able to navigate the taxiways and gates of existing airports.

One cannot talk about wings without mentioning winglets like those shown in Fig. 49 (page 123). Patent documentation claims that they save fuel because they effectively increase the wing's aspect ratio. They also cause the wing tip vortex that always accompanies lift to dissipate faster. The latter is good for traffic management in airports because landing aeroplanes can follow one another more closely with reduced concern for flight path disturbances associated with the vortices from an aeroplane ahead. How much is hard to say, but they certainly look good. Whoever claims the innovation as their own seems to have been and may still be in dispute with the major makers of aeroplanes. One may suppose that good ideas can come from various places and even simultaneously.

About Speed and Engine Installations

What does the normal operating speed have to say about the way that the engine is designed into the design of an aeroplane? The engine manufacturer builds the machinery as a package, consisting of compressor, combustor and turbine, together with connections for ancillary equipment like controls and electrical and hydraulic power systems, bleed air, etc. The aeroplane manufacturer has to integrate the engine into the airframe because the performance of the aeroplane together with its engines counts, particularly on fuel economy.

Integration into the aeroplane means that the aeroplane builder has to design the nozzle forming the jet and the inlet that admits the fresh air. The nozzle design is pretty straightforward: its area has to be fixed for the right jet speed, and noise emissions have to be considered. The inlet, on the other hand, is more complicated. Ideally for flight in cruise, the inlet should be nothing more than a sharp-edged scoop that captures air and slows it down a bit. A rather extreme

Fig. 73: Inlet on the nose of an F-100, top, and B-58 nacelle, right. These aeroplanes are designed to fly at supersonic speeds. (Top: US Air Force photo 021210-F-1234P-041; right: author at the National Museum of the US Air Force, Dayton, Ohio).

example is shown on the US Air Force F-100 Thunderbird and the B-58 inlets in the figure above.

The slowing down of subsonic flow is easily done with a duct that smoothly increases in flow area (picture a hand-held megaphone) and that slowing helps the fan or compressor blades from having to deal with the near sonic flow at the higher flight speed. The inlet also helps the compression function because the air pressure rises as the flow proceeds through it. That is the easy part of the design.

The hard part has to do with the requirement that the engine must receive nice and uniform airflow at low-speed conditions, take-off in particular. At low aeroplane speed, the air entering the inlet comes from nearly all directions, including from the region to the rear of the inlet lips. Ideally the inlet might want to look like the open end of a trumpet under these circumstances. Such a design would be absurd. The flow entering a realistic inlet has to negotiate the turn around the inlet lip and proceed smoothly into the inlet. If the inlet lip is too sharp, that nice smooth entry into the fan face does not take place and the engine has to deal with 'dirty' flow, some of it high speed and other parts nearly static. That situation can lead to stall of the fan and loss of thrust. That is very undesirable during take-off. Further, the aeroplane is, at times, subject to cross winds that can compound the problem.

One solution is to either admit more air from the back side of the inlet cowl with what are called *blow-in doors*, another is to design the inlet with a sufficiently rounded lip so that stall is avoided under all foreseeable conditions. The blow-in doors work well enough: they open when the pressure difference across the spring-loaded doors pushes them open and they close when in cruise flight when pressures are more balanced. When the doors are open, some of the air can bypass the need for flowing past the inlet lip.

Two things have evolved to render the blow-in door less than desirable. First is the energy crisis of 1973. Continued environmental concerns and fuel usage costs have reduced the flight Mach number for aeroplanes from near 0.90 to the 0.80s and lower. The reduction in flight Mach number allowed for a more rounded inlet lip. The second undesirable aspect is that the flows through the doors are separated by wakes from the structure between doors. That makes the fan flow significantly noisier than it would be if the flow were more uniform. Obviously, since the blow-in doors are used near take-off, it is the airport environment that suffers the noise impact. The JT3D and JT8D engines were commonly equipped with blow-in doors in the early years of their use. Such installations can be seen today on museum aeroplane articles like the Boeing 720 and 737. Boeing 747s also had blow-in doors on the early JT9D engines.

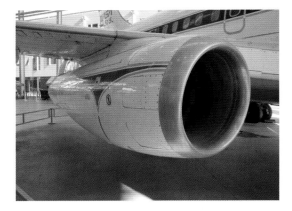

Fig. 74: Top: Blow-in doors on a 737-100. Three visible doors are closed on left image and internal air passages into the same inlet are shown on the right. Note that the engine mount does not allow for air to enter around all 360 degrees of the engine. (Photo by author at the Museum of Flight). Right: Blow-in door inlet on a Boeing 747. This aeroplane is climbing out from take-off with leading edge flaps still deployed. The engines are P&W JT9Ds. The photo was taken during an evaluation of the outboard inlet configuration without blow-in doors. (Boeing Images BI411506).

Towards Twin Jets

The successful launching of long-range air service with jets quickly led to airlines wanting to serve what is called the medium-range market where distances between cities are about 2,000 to 3,500 miles apart. For this market we have an interesting evolution history of the airliner design. To display that history we have to take a longer perspective that includes the Boeing 720 but also the latest 2018 versions of the 737. Consider the evolution of Boeing aeroplanes. Boeing has had the longest history and the most experience with dealing in the jet airliner market. All other American competitors have dropped out of the business of building complete airliners. The only competitor of any significance is the French–German consortium Airbus. Its history dates back to 1970 with an offering of the A300, the first twin-aisle, twin-engine aeroplane. Since then Airbus and Boeing have been actively engaged in reading the tea leaves of the airliner market with offerings of aircraft that are similar when viewed from the larger perspective but with differences targeted at specific customers. With that thought in mind, a look at Boeing's history is representative of the larger picture.

Consider that the Boeing 720 (date of service entry to end of production of last version: 1960 to 1979) was a variant of the 707 designed as a mid-range aeroplane. It had four engines of about 12,000 lbs thrust each. The replacement aeroplane, the 727 (1964–84), had three 14,500 lbs thrust engines. The next replacement, the 757 (1983–2004) had two engines of 36–43,000 lbs thrust. Finally, as the capability of the once smaller 737 and available engines grew, it went on to replace the 757 in the Boeing lineup.

Early in the history of the 737, the JT8D engines started at 14,000 lbs thrust. Today, in 2018, the aeroplane employs CFM International (CFM56 and CFMLEAP) high-bypass engines with twice the thrust. The aeroplane performance improvement that followed is evident in its wide use that today holds the record for sales: over 9,000! The road to such success was not always evident and therein lies a bit of interesting history told in the next chapter.

The trend is clearly evolving to twin-engine aeroplanes – and to think that the large jet aeroplane industry started with the B-47 with six engines. Going back further, we have the B-36 with ten engines, or the Hughes H-4 Hercules with its eight engines, the latter of these being the most powerful piston engines ever built!

The aeroplane manufacturing industry is trending in the direction of twins as long-range aeroplanes. For Boeing, consider that that the 707 (1958–79) and the 747 (1970–present) have four engines and were to have been outdone, albeit unsuccessfully, by three-engine aeroplanes. These were the Douglas DC-10 (1971–88) and the Lockheed L-1011 (1972–84). The design of these two aircraft was quite similar except for the centre engine configuration. The DC-10 had a straight-through mount in the vertical tail while the L-1011 had the engine at the rear of the fuselage with an S-shaped duct, not unlike that used in the Boeing 727, to provide air from above the fuselage. The L-1011 engine location required a short engine for installation there and neither the Pratt & Whitney JT9D nor the GE CF-6 would fit. Thus, Lockheed was stuck with the RB.211 and Rolls-Royce was stuck with Lockheed in their attempt to play a role in the US airliner market.

This last constraint had to do with foreign engines on American aeroplanes and the political implications on employment. Heppenheimer in his book *Turbulent Skies* tells the story of the wide-body trijets well and in greater detail than here. The fierce competition between these similar trijets had a lot to do with both producers' demises in the airliner business. By the time production ended in the mid to late 1980s, 386 DC-10s and 250 L-1011s were built. The basic design of the DC-10 saw a life extension to the year 2000, the end of the trijet era, as the McDonnell-Douglas MD-11 of which 200 units were sold.

Boeing considered a three-engine jet to compete in this arena but ultimately settled on a twin-engine design skipping the three-engine configuration altogether. Since its first flight in 1972 and service entry in 1974, the European consortium Airbus had shown the viability of a twin-aisle, twin-jet airliner. A significant number – 561 A300 (produced 1971–2007) and 255 of the smaller derivative A310 (1983–98) – proved to be effective competitors to the American airliners. The Boeing Company's competitive later offerings were the wide-body 767 (over 1,000 produced 1982–present) and the even larger 777 (over 1,500 since 1994). Interestingly, the wide-body 767 was designed at roughly the same time as the narrow-body, single-aisle 757. In a rather interesting move, Boeing designed the noses of these two aircraft, *i.e.* their cockpits, to be nearly identical, saving design and fabrication expense as well as allowing flight crews to service both aircraft with relative ease. A result was that the cockpit floor levels of these aeroplanes differed by a small step from the cabin floor. Today the 767 is solely manufactured for military customers. Commercial production ended with the introduction of the 787 around 2011.

There are still other designs that reinforce the conclusion that the twin-engine configuration for a modern jetliner is close to optimal. These include many aeroplanes built by Airbus (including over 1,400 (as of 2018) of the twin-aisle, twin-engine A330, since 1992), the Bombardier C-series, and others from Asia on the drawing board or in development, as just a few examples.

Noteworthy exceptions to twin-engine configurations from Airbus are the A340 and the A380, both with four engines. The A380 is a double-deck, mega-aeroplane that appears not to be as successful as initially hoped. It was designed to carry from 550 to over 800 passengers in 2+4+2 and 3+4+3 economy seating in the upper and lower decks respectively. In early 2019, it was announced that A380 production was to be discontinued. The prognosis for the four-engine 747 also seems precarious at this time. In 2017, United Airlines and Delta flew their last 747s in revenue service and no other American airline retains the passenger version of the 747 in its fleet. The 747 does however remain in long distance

Fig. 75: Widebody trijet centre engine installations: Lockheed L-1011 (top) and Douglas DC-10 (bottom). Both pictures were taken on the types' inaugural flight days: L-1011 at St Louis 25 June 1972 and DC-10 at Los Angeles, 17 August 1971. (Photos: Jon Proctor)

service with many foreign airlines. Production is currently not as vigorous as it once was, half a century ago, and a good fraction is the freighter model.

Not to be omitted in this story is mention of the world on the other side of the Iron Curtain. There, about 100 four-engine wide-body Ilyushin Il-86 airliners were built between 1976 and 1991 and retired from service about ten years thereafter. The Il-86 was powered by Kuznetsov NK-86 (bypass ratio ~1) turbofan engines. The follow-on and similar Il-96 is currently still in production, at a low rate. It was introduced in 1992 is powered by Russian and, interestingly, Pratt & Whitney PW2000-family engines.

The high reliability of modern engines obviates the needs for four engines as a means of insuring flight safety. Aviation officials give twin-engine aircraft a rating[12] that allows them to operate over the long distances of the world's oceans.

Thrust and Aeroplane Weight

As one looks over the jet engine thrust of airliners over the years, it becomes apparent that the relationship between thrust and aeroplane weight is quite well correlated. For example, the sea-level, static thrust of the engines of the first Boeing 720s was about twenty per cent of the take-off gross weight. With the switch to turbofan engines, this percentage jumps to nearly thirty. If one looks at all the airliners built since then, that percentage ranges between 25 and 30. The higher percentages result partly from the fact that most modern aeroplanes have two engines rather than three or four. It is a requirement of certification authorities that an airliner must be able to fly under all conditions with one engine not functioning. In a two-engine aeroplane, one engine lost means that half the normal thrust would be available. By contrast a one-engine-out situation on a four-engine aeroplane means a loss of only one quarter. This aspect of aeroplane design accounts for the rather lively performance of a twin-engine aeroplane on take-off. The acceleration, g's if you will, felt by the passengers when the brakes are released on the runway is a direct measure of the aeroplane's thrust to weight ratio.

In short, the engine thrust available dictates the weight allowable for the aeroplane. Furthermore, the higher the thrust available from an engine, the more realistic is a twin-engine design.

---

12. The rating is the so-called ETOPS rating, allowing aircraft to fly over regions more than a specified number of hours away from an airport.

Thrust and Engine Weight

There are many ways to characterize the progress made over the years in jet engine performance. The engineer might cite efficiencies of components: the compressor, the fan, the turbine, etc. Also of great interest is the effective turbine inlet temperature. There is no doubt that these parameters have improved dramatically since 1945, but getting solid numbers poses challenges to being accurate and meaningful. Similarly, fuel consumption has improved with the employment of the technologies described here, and comparable numbers are also hard to sleuth out; such detailed performance parameters tend to be held closely by the engine manufacturers. There is, however, one parameter that is fairly easy to find: the thrust to weight ratio of the engines. The simple turbojets of the post-war period weighed about as much as the sea-level static thrust they could produce. The table below shows the dramatic improvement that has been made since then. Thrust to weight for a jet engine is somewhat akin to horsepower per pound for a reciprocating engine (although not quite comparable). Below we list the jet engine's evolution in these terms separately. The thrust level produced has grown for the engines in this listing by almost two orders of magnitude, from 2,000 to over 100,000 lbs, speaking clearly to the increased performance of our modern engines. The choices in the table are meant to be representative of the technologies and not a technology review of all the engines that participated in this history.

Engine	Thrust/weight
Jumo 004B	1.19
Rolls-Royce Derwent I	2.05
GE J47-GE-25	2.34
Pratt & Whitney J57-P-23	2.26
Pratt & Whitney JT3D-8A	3.69
Pratt & Whitney JT8D-219	4.43
Pratt & Whitney JT9D	5.81
GE CF6-50	6.07
GE90-115	5.98

Table 2: *Thrust to engine weight ratio for gas turbine engines over their history. Note the slight decrease represented by introduction of two spools in the J57 turbojet over the simpler J47 and the improvement brought about by the introduction of the low bypass turbofan engines, the JT3D and JT8D.*

It may be tempting to compare the performance of jet engines to that of piston engines from the viewpoint of engine weight. Such a comparison is made difficult because the appropriate measures of performance are, respectively, lbs of thrust and horsepower which are not the same measure. A rough comparison may nonetheless be made by taking the power in the primary and fan jets of a high-bypass turbofan at static take-off conditions and relating that to the engine weight. A number like 5–6 HP per pound of engine weight results for the gas turbine engine powerplant. This advantage of the gas turbine over piston engines is similar to that enjoyed by turboprops (that actually produce horsepower) over the piston engines during their last hurrah when the latter just managed to weigh in at one horsepower per pound of engine weight.

Chapter 8

How to Buy an Airliner

How does an airline executive decide whether to buy a few airliners that an airframe builder might offer? The airline official will have to consider purchase price, fuel and crew expenses, airport fees, maintenance and other costs to be balanced against what the airliner can earn in its hopefully long life. Fuel consumption rate is a cost component that an aeroplane or engine designer can and must have a good handle on. The price of fuel is another matter. The airline executive will probably ask: 'Will you guarantee the fuel use performance of your aeroplane when I put the aeroplane in service?' The answer is, of course, 'Yes!'

The aeroplane manufacturer must have a pretty good idea about what he can promise and guarantee. The issue is thorny: a very safe guarantee level from the manufacturer will result in sales lost to a competitor who may be willing to take more risk by guaranteeing a better performance level. If the manufacturer strays over the real performance level by promising something better than he can deliver, he will be contractually obliged to pay for the excess fuel used, or something equivalent to that. When you have a fleet of airliners operating sixteen hours a day every day of the year, a small error can quickly devolve into a financial nightmare.

Tools of the Aeroplane Designer

In the jet age, the modern-day computer also came of age. A considerable treasure and expertise was spent in the last decades of the twentieth century to develop computer capacity, to describe the physics of airflow in mathematical terms and create a virtual model of an aeroplane whose performance could be assessed. That was the ideal world. Unfortunately, real life is more complicated. The physics of airflow is notoriously difficult to predict, much less quantify, to a high degree of accuracy. We would desire an accurate performance assessment of lift and drag under all flight conditions, most importantly under cruise conditions where an airliner spends a large fraction of its time, and a detailed knowledge to comfort the engineer that alterations to routine conditions through weather, pilot idiosyncrasies, etc, can be easily and safely handled. The computer today is very capable. The equations that are built into it for performance analysis are in pretty good shape and improving. They remain imperfect however. Friction effects, for example, can render the air flow field unpredictable to a sufficiently high level of accuracy. Fortunately this is the case only in some parts of an aeroplane, under some conditions. Newton's laws apply easily to places where friction is absent and there the mathematical description is relatively easy. They also apply to places where friction is important, but the equations are much more complicated and the boundary between these regions is tenuous at best because their location is a challenge to properly describe.

In regions where friction is absent, the analysis of the flow is straightforward but, and this is a big one, not necessarily so if waves are present. When we fly at speeds near the speed of sound, these waves are unavoidable and their influence reaches into the flow field where friction is important and may play havoc there by creating a situation that involves drag in amounts that are a challenge to determine accurately.

For example, aeroplanes routinely fly at Mach numbers high enough so that locally supersonic flow is experienced over the wing. A supersonic flow jumps back to the subsonic flow velocity by means of a shock wave that forms near and

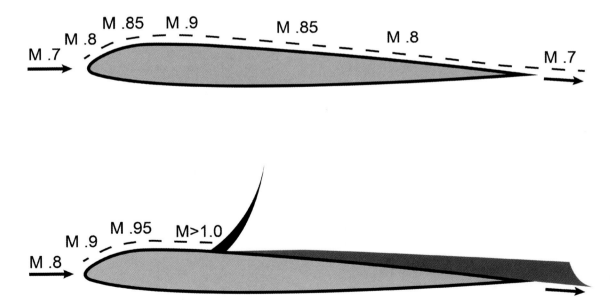

Fig. 76: Streamline pattern over an aerofoil at subsonic speeds (top) showing the flight speed variation as Mach number over it. The bottom at transonic speed leads to a shock wave near the point of maximum thickness and potentially separated flow due to the shock-boundary layer interaction. The pressure over the aft portion of the aerofoil is lower than it would be without the wake formation (sketched in dark grey), hence a drag increase is experienced.

ahead of the wing's trailing edge. As long as that wave is weak, there is no problem because the consequences of that wave have been used to specify the cruise Mach number that the aeroplane should fly with good economy. Flying faster, however, makes the recompression shock wave stronger. That increases the likelihood that the flow will detach from the wing's upper surface and create a thick wake that is costly in terms of drag. The formation of a wake also occurs when the wing approaches stall and then suffers a drag increase

Vortex Generators

Sometimes, even in the absence of shock waves or high angles of attack, the flow routinely separates from the surface it is hoped to have followed. In that case, small devices can be employed that cause an exchange between high-energy airflow obtained 'far' from the surface and air with reduced energy or speed near the surface. The low energy air near the surface has been affected by surface friction and is referred to as 'boundary layer' air. The devices in question create vortices that allow for that air exchange. They are permanently mounted where they are needed and are called *vortex generators*. They have been used since their invention[1] in the early jet years by an engineer at the United Aircraft Corp where Pratt & Whitney is a corporate family member. Such vortex generators are easy to spot on museum displays of entire aircraft. Fig. 77 shows the vortex generators on the wing upper surface above the inboard engine of a Boeing 707.

Air flow vortices also play a role in the design of very modern aircraft at the exhaust edge of propulsion nozzles[2] to suppress noise while minimizing the performance losses associated with the mixing of jet and free air stream flows. Examples of two patents relevant to the design of a modern turbofan engine nacelle are given below to show the general time frame of their discovery and the corporate entities involved. Note that 3 to 5 years pass between patent application and official publication.

---

1. Cook references the work of H. Bruynes and H.D. Taylor, 'Fluid Mixing Device', US *Patent*, 2,558,816 (3 July 1951, filed 16 August 1947) at the United Aircraft Research Department in connection with high speed pitch up wing tip stall on the B-47. See also 'Vortex Generators', US Patent 2740596A, by John G. Lee, 3 April 1956.
2. 'Chevron Exhaust Nozzle for Gas Turbine Engine', US Patent 6360528 B1, by John F. Brausch et al, (General Electric Co) 3 March 2002, and 'Gas Turbine Nozzle Configuration', US Patent 8358468 B2, by David A. Cerra et al, (The Boeing Company) 22 January 2013.

Fig. 77: Vortex generators on the wing above the inboard engine. The engines are JT3Cs on an American Airlines Boeing 707-123 showing the noise suppression daisy petal mixer on the outboard engine. (Photo: Jon Proctor).

Fig. 78: Propulsion nozzle of a modern engine nacelle (Boeing 787 with Rolls Royce Trent engines). Note the scalloped nozzle exit lip acting as built-in vortex generators. (Photo by author at the Museum of Flight, Seattle.)

Wind Tunnel Testing

A historically important way to get around the difficulties of accurate drag determination was and continues to be the traditional method of wind tunnel testing. Realistically, the scale of the model in the wind tunnel cannot be the same as the article being investigated. Thus, mathematical tools to estimate the full-scale aeroplane drag from small-scale wind tunnel model data require careful analysis. This scale difference is notoriously difficult to bring to bear on the problem of predicting the drag performance of an aeroplane to a high degree of accuracy.

Internal Engine Flows

The issues associated with lift and drag forces on an aeroplane are also present in the performance prediction of the engines. Propulsion engineers work with moving air bounded by walls and the physics is quite similar. For the internal flows of an engine, the uncertainties of flow calculation are buried in parameters that quantify an engine's component performance. The important ones are the compressor and fan efficiencies as well as similar measures for the turbine(s). These efficiencies measure the proximity to performance perfection. While these typically proprietary efficiencies are nominally between eighty and ninety percent for compression and somewhat higher for the expansion in a turbine, they describe how much, or better how little, of the power involved is converted to less useful heat. The engine designers obviously try their best to maximize performance. When these efficiencies are known accurately and the temperature of the combustor outlet is known, the prediction of an engine performance is relatively easy. Only air bleeds and mechanical power removal for various purposes constitute additional performance issues.

Both the engine manufacturer and the aeroplane builder have accuracy uncertainties to deal with. It is important for the engineer to know just how big the uncertainties are. Is it one percent or is it ten percent? To be on the safe side, the aeroplane builder might have to give up a lot of performance promise if the number is anywhere near ten percent. The uncertainties are confounded by interactions between the flow-field around the aeroplane associated with the engine. In this sense, the integration of the engines on the aeroplane is yet another element of performance uncertainty. A number of examples from Boeing history may illustrate the point.

Engine-Airframe Integration

Consider first the Boeing 727. The rear mounted engines, especially the centre engine with its S-duct, were a novelty at the time this aeroplane was designed. An attempt was made to understand the flow field with various kinds of wind tunnel models that could not include a scale model of a jet engine. Ideally the goal would have been to model the jet and the inlet into an engine simultaneously. This was approximated with 'simulations' that included a hollow tube to model inlet airflow of an engine and another model with compressed air introduced into the model. In the second test the model had a closed inlet and the results yielded an estimate of the 'jet effect' on aeroplane model drag. The results provided information about the net effect of the engine installation so that performance guarantees could be written for the sales department. They were, however, just good estimates. The real data had to await construction and flight test of the real thing. It turned out that the drag performance of the 727 in flight was significantly better than predicted, so better guarantees could be written. A quick memo to the sales department eased their task considerably. In time the 727 was very successful with sales of 1,832 aircraft in the years between service entry in 1964 and the following twenty years when the 757 was introduced to replace it.

The 727 also encountered a problem with the engine installation in that the centre engine was initially prone to stall. On take-off rotation this is a fairly serious problem. It turned out that the S-duct inlet flow did not provide a sufficiently 'clean' (uniform) flow at the engine inlet face. Fortunately the application of vortex generators in the S-duct fixed the issue. The 727 is a good story; now for one which was less happy, at least initially.

Our second case deals with the 737. When this aeroplane was introduced into service in 1968, it was equipped with JT8D engines mounted on the wing. It was a small aeroplane for eighty-five passengers to fly relatively short routes. The short landing gear forced the engine installation under the wing to be quite close. The unforeseen consequence was a strong engine-wing flow field interaction that resulted in drag exceeding performance estimates. Boeing recognized the problem and reconfigured the engine nacelle to alleviate it. Increasing the landing gear height was not practical because changing the wheel well and/or strut location on the underside of the wing would have been enormously costly. The fix involved relocating the engine forward of the wing on a slim nacelle strut. This was also very expensive – fixing an aeroplane in production is always costlier than designing it correctly in the first place. The cost was so high that engineers of the day thought that the 737 would never make a dime for the company. The story ends more happily in that not only was the performance guarantee met with a new engine

Fig. 79: Boeing 737 Engine nacelles on a 737-200 for a JT8D (left, service entry 1968) and a CFM56 high-bypass engine on a 737-800 (right, mid-1990s). Note the external hinge points for the clam shell thrust reverser on the rear of the -200 engine oriented to direct the reverse jet over the aeroplane body and to the side the aeroplane to avoid debris ingestion. Note also the 'hamster' cheeks shape of the CFM fan cowl (on the right) where various engine components are housed for sufficient ground clearance. The -800 aeroplane is headed for the paint shop. (Wikipedia Commons and Michael Carter – Pacific Aero Images).

installation, but the aeroplane did well in the airliner marketplace against the narrower-body DC-9, its chief competition. Later models of the 737 with CFM56 engines improved the aeroplane's capability to the point where the 737 became the most produced aeroplane in airliner history. The CFM56 engine is a high-bypass engine that again challenged the designers: the large fan diameter posed ground clearance concerns. The solution was a configuration where the engine accessories are located at the 4 and 8 o'clock positions rather than at the 6 o'clock bottom of the engine. Viewed from the front, the engine invited comparison to the full cheeks of a hamster! As of September 2017, a total 9,716 of the aircraft have been sold and it remains in production with a solid backlog of orders.

The switch in the early 1980s from JT8Ds to CFM56s on the 737 (-300 and later models) had an important impact on the noise produced. The effectiveness of noise reduction resulting from the use of high-bypass engines was surely a component of the decision to design a replacement for the 727 that was also powered by JT8Ds and where the use of high-bypass engines would have involved a costly redesign of the aeroplane. Hence the focus went to construction of a new twinjet, the 757. It entered service in 1983. The engines for the 757 were Rolls-Royce RB.211 and Pratt & Whitney PW2000s. Both were high-bypass engines with thrust in the 37–43,000 lbs range. The Pratt & Whitney appellation of this engine was initially JT10D but it is now named according to a new scheme.

The 727 and 737 as design examples have primarily to do with the necessity of understanding the engine-airframe integration and its effect on performance. A third example has to do with performance deficits on either side of the engine-airframe divide. It is a challenge to discern just who is at fault when performance data is worse than anticipated. Is it the engine's poor thrust performance or is it the drag on the aeroplane? Neither thrust nor drag can be measured in flight directly. They have to be inferred from complex calculations with parameters that have uncertainties associated with them. No matter where the problem resides, thrust and drag are always in balance in level flight. When the 747 prototype first flew in 1969 there was difficulty in determining whether the airframe or the engines were responsible for worse than anticipated performance. In the end it was found that the engines needed to be reworked for a variety of reasons, some of which were cited earlier. Since 747 production was underway, with many customers lined up to fly them, Boeing produced a number of aeroplanes without engines. These were stored for a while with concrete blocks in place of the engines, waiting for the new improved ones. The concrete blocks ensured that the aeroplane's centre of gravity rested firmly between the wheels. Fortunately that story also turned out well for Boeing and Pratt & Whitney and history speaks clearly of the 747's success.

737 MAX Accidents

To close out the discussion of engine airframe integration for airliners, we note that airframe design includes more than just the physical parameters of the configuration and their effect on drag. In the case of the latest Boeing 737, the operational consequences of configuration changes to aeroplanes already in service turned out to have serious consequences.

In late 2018 and early 2019, two fatal accidents of Boeing 737 MAXs are again calling attention to the importance of engine-airframe integration. These apparently similar accidents involved an uncontrolled descent relatively soon after take-off. The loss of 346 lives caused the affected 737 fleet to be grounded pending completion of an investigation of the issues involved and their resolution.

The accidents took place in the context of fierce competition for the single-aisle, twin-jet, short to medium range airliner market. While a more detailed description of the market evolution and its breadth is beyond this writing, a major element is the dominant positions of The Boeing Company's 737 and the similar Airbus A320 together with its variants A319 and A321. The reader may wish to obtain more precise numbers but the reality is that both Boeing and Airbus have delivered many of these aircraft, with orders for thousands more on the books. The production and order numbers for the two companies are quite similar. There were and are other participants in that market, with aircraft and engines discussed here and elsewhere, but it appears that it was competition between the topmost producers that played a role in the accidents.

Boeing recognized early (1980) that the 737 with JT8D engines was going to require better fuel consumption and lower noise, and in the mid-1980s introduced high-bypass CFM56 engines that saw service in the classic 737-300 to -500 models and the New Generation 737-600 to -900 models. The A320 was introduced somewhat later (1988) also featuring high-bypass engines: the GE and Safran built CFM56, V-2500 engines produced by IAE, and Pratt & Whitney PW6000s.

The 737's CFM engines had a bypass ratio of 5.1 to 5.5 and fan diameters of 60–61 inches. The decision to use the fuselage of the early versions of the 737 with its short main landing gear forced the CFM engine nacelles to be modified specifically for the 737. Thus the CFM56-3 (737-300 to -500) and -7 (737-600 to -900) versions of these engines were configured for sufficient ground clearance with their distinctive non-circular inlet (see Fig. 79).

The next step in the evolution of these aeroplanes was led by Airbus with the introduction of the A320neo (new engine option) family, again including stretched and smaller models. The programme was launched in December 2010. The engines available had a very high bypass ratio for greater improvement in fuel consumption. Specifically, they were the CFM International LEAP (Leading Edge Aviation Propulsion!) engines (fan diameter 81 inches, bypass ratio of 11) manufactured by the consortium consisting of General Electric and the French company Safran. Also available was the geared fan Pratt & Whitney PW1100G engine (78 inch fan diameter and bypass ratio of 12.5). The fact that the A320 is an original design allowed it to readily accommodate the higher bypass ratio engines with larger fan diameters. The A320neo went into service in 2016.

Boeing responded with an upgrade of its New Generation line (737-600 to -900) also to improve fuel use performance and again to avoid the costs of producing a new aeroplane. The decision was made in 2011 to offer a modified 737, called the 737 MAX, in a number of versions with various seating capacities and commensurate airframe weights. The MAX family maintains use of the standard fuselage, improved with a number of changes. The most important improvement to the aeroplane was use of the LEAP engines with a bypass ratio of 9 and fan diameter of 69 inches. The bypass ratio and fan diameter of these LEAP engines were tailored specifically for the 737 MAX and are smaller than those used for the A320 because of limitations imposed by the decision to continue use of the original, low to the ground, 737 fuselage.

The larger engines required their location to be further forward and higher on the 737 wing. The larger diameter also necessitated the nose gear having to be made longer to ensure sufficient ground clearance for the larger engine. The higher thrust and the new engine installation changed the flight characteristics of the aeroplane when compared to the earlier 737. Boeing also felt strongly that pilots trained on the older 737 must be able to transition to the MAX without additional training. Thus the changed flight characteristics demanded that steps be taken in the design of the control system to provide 737 pilots an easy transition to this newer aeroplane. It turns out that the engine installation required a means to avoid a flight stall that would more readily occur than with the earlier models of the 737. The solution spelled out in the certification process for the MAX by the US Federal Aviation Administration was to install an automatic control to override the pilot in order to avoid a stall in flight, especially in climb conditions.

The 737 MAX entered service in May 2017. In the marketplace the aeroplane family performed very well, lining up orders to make this aeroplane the most productive ever for Boeing. Several hundred new MAX aeroplanes were delivered by the time the accidents occurred. The mandated control system and arguments about whether it required supplemental pilot training seem, in March of 2019, to be at the heart of the two accidents. More cannot be said until more facts come to light.

Fig. 80: Front and side views of an unpainted Boeing 737 MAX engine nacelle showing its location on the wing. Photo by author at the Boeing plant in Renton, Washington.

While this writing is not meant to be reportage of current events, it sometimes happens that history is made while historians document history. The important point however is that engine-airframe integration is difficult and the challenge associated with the 737 MAX will likely be met successfully, albeit at some cost to Boeing. It has to be, because the economic and reputational consequences of having a grounded fleet of new aeroplanes are serious.

Considering the consequences of Boeing's decision to build the 737 MAX by modifying an existing aeroplane, one may wonder whether the costlier approach of designing a new wing incorporating a higher main landing gear might not have been a wiser path. Another practical option might have been to resuscitate and adapt the proven 757 that had served the company rather well.

The topic of determining and optimizing aeroplane configurations can be closed with a question about the future: With an ever increasing-capability of computational fluid dynamics to establish the performance of flight configurations and a convergence of the designs of airliners towards twin-engine aircraft, will wind tunnels ever again play the role they once did in the design of commercial aeroplanes?

Aeroplane Weights and Range

A number of parameters have been mentioned with some of their numerical values. An engineer cannot get away from these and these last paragraphs are to pull together everything that he or she would need to judge the viability of an airliner design. The central concern is the question: How far will a given fuel load take the aeroplane?

We begin with the typical weight numbers that have to do with airliner operations. First and foremost is the maximum weight that can be launched safely. The engineer calls that the *maximum take-off gross weight* (TOGW), sometimes called the brake-release gross weight. For the early models of small, short range aeroplanes like the 737 or the DC-9, this might be about 100,000 lbs. For the 747, this weight ranges from 700,000 to one million lbs, well exceeded by the Airbus A380 at 1.27 million lbs.

There are other important weights associated with the operation of an airliner. As an example, consider the 747-8 with a TOGW of 987,000 lbs, which we will call 100% to make the discussion easy and more general. One such weight is the maximum landing weight. The stresses associated with landing are potentially large enough that designing the landing gear to be sufficiently strong to withstand the landing impact of a fully loaded aeroplane would be unrealistic. Further, the need to land fully loaded is extremely rare, so why plan for it? If an airliner has to land shortly after take-off, the pragmatic solution is to dump excess weight (fuel) before landing. In the case of the 747-8, the maximum landing weight is 688,000 lbs (~70%).

Other important weights are the *maximum zero fuel weight* (ZFW) and the *operating empty weight* (OEW). The ZFW is the weight of a fully loaded aeroplane without fuel, or 651,000 lbs (66%) that must be smaller than the maximum landing weight. The OEW is the aeroplane weight without payload (passengers and freight) and no fuel. It is the bare aeroplane weight, 485,000 lbs (49%). The maximum allowable payload weight is the difference between the ZFW and OEW, 17% in this case.

Another interesting weight is the maximum *fuel capacity weight*. The fuel tank capacity installed allows just so many pounds to be loaded, 426,000 lbs (43%). Fuelling the aeroplane to this capacity is useful when very long range is desired with a limited payload. The payload for a long-range trip with a full fuel load is TOGW minus OEW minus fuel weight, or 8% in this case.

The range achievable by any aeroplane is determined by the take-off gross weight and the landing weight. For a fully loaded aeroplane with just enough fuel to stay within the take-off weight limit, technical specifications show this range to be about 6,800 (statute) miles for the 747-8.

This writing is about propulsion so that a mention of the engines' contribution to the aeroplane's weight is in order. For the 747-8 the four GE engines weigh in at roughly 50,000 lbs. Thus, engine weight is roughly five per cent of the TOGW fraction or ten per cent of the airframe weight, *i.e.*, OEW.

The Range Equation

The range equation is a mathematical statement that aggregates the various design aspects of aeroplane and engine designs and determines the success of operating an aeroplane in the real world. The range, given in a technical document from the aeroplane manufacturer, always includes all flight elements from brake release to landing. These elements include climbing to and descent from altitude, managed to minimize fuel consumption. For a long-range aeroplane, one can look at an aeroplane in steady level flight and ask what are the parameters that determine how far it can fly with weight always balanced by lift and drag always balanced by engine thrust.

The engine consumes so many pounds of fuel per hour for each pound force of thrust it generates. This number is the *thrust specific fuel consumption*, or *TSFC*. It is measured in lbs of fuel per lb of thrust per hour in American engineering circles.

A little calculus yields the very simple result[3] that the range is directly proportional to the flight Mach number, *M,* and the aerodynamic design of the aeroplane as described by the lift to drag ratio, *L/D*. It is also inversely proportional to the engine performance through the thrust specific fuel consumption, *TSCF*. The values of *L/D* and *TSFC* are the sum total of the performance indices for the aeroplane and for the engine, respectively.

---

3. Range, in statute miles = $660 * M * L/D * 1/TSFC * \ln(TOGW/ZFW)$

The technically orientated reader can look up the range equation and discover that it contains a term having to do with the weights of the aeroplane at the beginning and at the end of the flight. The weights are determined by the way the aeroplane is operated.

We again examine the 747-8 as an example. Let's see what the range would be if the aeroplane was only in cruise without considering climb and descent. The TOGW is 987,000 lbs. It would be somewhat lighter at the start of actual cruise. The weight at the end of mission would be larger than the maximum ZFW to allow for safety reserves, but let's use these numbers. For these numbers, the *ln* of the weight ratio term in the range equation is 0.416. An estimate of *L/D* may be 18 and *TSCF* may be 0.6 lbs of fuel per hour per pound of thrust at altitude, and we obtain a range of 8,200 miles rather than the quoted 6,800 miles for a realistic flight. The difference in these two distances reflects our simple view of the issue in relation to a more realistic calculation. In an accurate determination of the range, the weight aspects of this calculation are accounted for by close monitoring of the fuel usage during true cruise conditions. Under such circumstances, lift and weight are balanced, as are thrust and drag. These balances are not achieved in climb and acceleration.

At the heart of distinguishing between the performance levels of engine and airframe lies the troubling fact that in the range equation the *L/D* and *TSFC* appear in a way that makes it impossible to tell which one might be responsible for a range shortfall. The centrality of these parameters to the determination range is part of the reason that manufacturers tend to keep precise knowledge of these numbers confidential. Furthermore, neither thrust nor drag can be measured directly. They must be calculated for the determination of *L/D* and *TSFC*. Thrust is determined from computer calculations that contain the best estimates for the many details that govern overall engine performance. Under the best of circumstances, both aeroplane and engine manufacturer agree about the component performance parameters. Similarly, drag is also calculated, typically by the aeroplane manufacturer's aerodynamics staff.

Two more aspects are interesting. First, we note that modern aeroplanes like the Boeing 787 wherein a large-scale substitution of composites for metal was made in the design, the reduced empty weight translates directly into weight that can be transported as payload. Hence we have the motivation for implementing practical means to reduce the OEW of aeroplanes.

The second aspect concerns *L/D*. The lift-to-drag ratio is fairly constant in value as flight Mach number is increased; that is, until the transonic flow effects on drag become initially noticeable and then severe. The range equation states that the *product* of flight Mach number, M, and L/D should be maximized. Indeed, that is close to the flight Mach number that a specific airliner will have to fly to do it most economically. As an example, consider that L/D for a 727-200 decreases from 15.5 to about 12 as cruise Mach number is raised from 0.76 to 0.86, showing how sensitive L/D is to flight Mach number. Cruise near $M = 0.80$ gives best range or maximum $M \star L/D$. The *TSFC* also varies with flight Mach number, but the variation is weak for engines used in subsonic airliners. Hence engine considerations for choosing the best cruise flight Mach are minimal. Parenthetically, we note that we use here the data from the Boeing 727 because it is old enough to be readily available in the public literature domain.

Cruise Altitude

One last comment regarding the quantitative dimensions of flight is worthy of mention. As an airliner starts the cruise leg of a journey, it is heavy with fuel. Near the operating point of optimum L/D, the lift (and the equal weight) of a particular aeroplane's wing depends primarily on the local atmospheric pressure and the flight Mach number. The Mach number is chosen to be for maximum $M \star L/D$ as argued above so that its value is fixed. That leaves only the atmospheric pressure to vary the changing lift. The consequence of the direct relationship between lift and pressure is that as fuel is burned off, the airliner will want to fly at higher altitude where the pressure is lower. Thus, the flight will start at what the pilot may announce to be the *initial cruise altitude*. He or she will then cruise ever

Fig. 81: An Air Canada Boeing 777-200LR taking off. This aeroplane, in its external design, is not unlike that of a number of twin-jets. The aeroplanes relatively easily meet the needs of airliners in various sizes. Details nevertheless differentiate these beautiful aeroplanes. (Photo Michael Carter – Aero Pacific Images).

higher as fuel is burned off, consistent with the 'lanes' of altitude available for the purposes of air traffic management. These lanes are typically 4,000 feet apart with segments of 1,000 feet dedicated to traffic in the four cardinal directions.

Interestingly, the Concorde cruised at altitudes higher than commercial subsonic jets and was therefore almost alone in occupying its region of the atmosphere. It consequently could and did climb in altitude exactly as demanded by the decreasing aeroplane weight without regard to traffic altitude restrictions.

Supersonic?

According to Newton, a high-speed aeroplane requires production of a jet with a velocity at least as high as the flight speed. That is the challenge presented to the engine designer and invariably involves afterburning in a turbojet engine and the fuel use penalty is quantified as a large value of *TSFC*. An additional challenge for a supersonic airliner is that it will legally be required to fly at subsonic speeds over land, especially over inhabited areas, because of the sonic boom pressure wave. That means that the engines have to be designed to be efficient performers with sufficiently low *TSFC* at low as well as high speed. That problem remains an unsolved challenge.

The aerodynamicist, the engineer who has to configure the shape of the aeroplane, faces a challenge of his own in that drag at speeds beyond the speed of sound rises significantly with increasing speed. In technical terms, when the flight Mach number of a well-designed supersonic aeroplane is substantially larger that one, the *L/D* of the airframe would have to be such that $M \star L/D$ is roughly similar to that of a modern subsonic airliner. Unfortunately an *L/D* that good has never been achieved on any configuration of aeroplane designed to fly supersonically and thereby achieve competitive airline economics with conventional subsonic aircraft.

These technical issues dominate the question of whether supersonic flight is practical. They were central to the operation of the only supersonic transport ever put into commercial service, the British-French Concorde. It flew at twice the speed of sound (Mach 2), primarily between New York and Washington in North America and Paris and London in Europe. The limited market was served by fourteen aircraft that were rather modest in size, about one hundred passengers in four abreast seating. As with all advanced aviation innovations, government funding was required to develop the Concorde. Further, the economics associated with the high rate of fuel consumption of the afterburning engines involved operation subsidies in addition to high ticket prices. In the end, the service life of the Concorde after introduction in 1976 was terminated in 2003. It is a great tribute to the visionaries and engineers who built this beautiful aeroplane that, in a limited sense, they succeeded.

Operational aspects also dominated a later attempt in the United States to build a more successful supersonic airliner. To that end, the Boeing Supersonic Transport the Model 2707, together with a follow-on project, was designed to be made large, with about 300 seats, and fast. The flight speed was designed to be Mach 2.7, where structural materials with high temperature capability would have to be used. Titanium would be used to replace the aluminum used for conventional aircraft. Propulsion requirements necessarily involved afterburning turbojets, as used in the Concorde. The development of this aeroplane would also require funding by the US government, because of the technical challenges and concerns about the commercial viability. After years of development, the project funding was turned down by the US Congress in 1971, partially for reasons associated with environmental concerns. The American SST was not to be. Bigger and faster were likely better, but not good enough for a purely commercial venture.

The Old Days in Jetliner Travel

While this book has been about the external design of the airliner, there is one interesting facet about the inside of the passenger cabin. The present-day internal configuration of the cabin did not arise with the beginning of the jet age. Aside from the seating, the first figure of this book shows the dominant feature of the cabin: overhead stowage bins for suitcases and other items. The early jetliners had instead open 'hat racks'. In the jetliners, they were typically rigid shelves for the storage of a variety of items that did not include things heavy enough to injure a passenger in the case of an encounter with air turbulence. The shelves served to house drop-down oxygen masks and signs to advise the passenger about seat belts and smoking. The hat racks were just that, a place to stow the hat, overcoat, or blankets. In earlier times, travellers dressed more formally than we do today because air travel was much more of an occasion. The racks had been installed for that purpose since the beginning of air travel, when they were often little more than a net with an elastic edge.

In the early 1970s, Boeing introduced the overhead storage bins we know today. The economic downturn at the time caused the company to look into ways to promote sales. The company accomplished the sales objective, but whether it was because of the overhead storage is debatable. Overhead storage is now, however, an integral part of the cabin interior and part of the processes of loading and unloading an aeroplane at the terminal. An older reader might enjoy a look at a cabin interior of an earlier time (Fig. 51, page 128) when jet air travel was … well, less mature and smoking was a much larger part of everyday life.

Epilogue

The central story of aviation after demonstration of the feasibility of controlled flight is power, power to go faster. Initially it was provided by the internal combustion engine for automobiles and adapted for flight. The first decade of the twentieth century saw power levels in the tens of horsepower. That quickly grew to one hundred, and then to two hundred in the decade that included the First World War. The best way to obtain such power involved a debate of whether air- or water-cooling was best.

Air cooling attempts include the use of rotary engines that in practice could not be built for more than two hundred horsepower. That approach was a dead end.

The cooling question remained open for the next three decades with the development of air-cooled radial engines and water-cooled V or inline engines. These engines were built to produce 500 HP in the mid-1920s, 1,000 HP in the mid-1930s, and finally over 3,000 HP in the 1940s and '50s. Both cooling methods were applied to military and commercial aircraft engines, but the power level was at a dead end, again. Breaking the 4,000 HP barrier was difficult and close to impossible in commercial practice.

The situation was saved by overcoming the limitation of air handling in piston engines and turning to a new type of engine, the gas turbine, that can handle much more air. Thus, as piston engines were working towards new maxima, the gas turbine or jet engine, in its infancy, was producing 2,000 lbs of thrust to propel aeroplanes during the war years. In this arena there was also a competition about how to design the best jet engine. The choice was between the radial flow compressor and the axial flow device for the same purpose. The radial compressor enjoyed an early practicality but was soon abandoned for the superior performance potential of the axial flow engine. While the debate was still raging, engines with thrust capability of about 6,000 lbs found wide applications. The Jet Age was born. Axial flow compressor engines were successfully devised in the period shortly after the war and their potential led to the almost complete extinction of radial flow compressor engines. Another dead end.

The gas turbine with axial flow compressors gained rapidly as the technology approached a mature state. The engines were simple, beautiful, and powerful, reaching 10,000 lbs of thrust and more in the military setting. The original patent idea of Frank Whittle to build a fan was put into practice and the industry has reached nirvana after the 1960s with thrust capability of 25,000, 50,000 and more than 100,000 lbs of thrust from engines that are very fuel efficient, reliable, and long-lived. The industry doubled back a bit, to employ a fan that is somewhat like a propeller!

So the story of aviation propulsion reaches a satisfactory state, but engineers are never completely satisfied; they will always push the edges of what seems practical and will always encounter bumps in the road.

Appendix

How Does the Jet Engine Work?

Fig. 82: Isaac Newton (1643–1727).

Any propulsion system working in a fluid (air or water) environment must be able to generate momentum, *i.e.*, cause the processed fluid to have a higher speed in the direction opposite to the direction of a desired force. That is the truth stated by Isaac Newton in his second law of motion. The moving fluid is the jet common to all propulsion systems. Even fish and squid exploit it and the reader can make a jet with a boat paddle in the water.

A flying aeroplane generates two forces. By pushing air downward, a wing generates lift upward, and a jet rearward creates a forward thrust. Birds create lift and thrust simultaneously in ways we may be challenged to fully understand and we would be hard pressed to replicate!

The aeroplane propeller forms a jet mechanically by acting on still air or air moving towards it. The jet engine creates it with an internal flow out of a chamber that continuously provides pressurized air that can be directed by means of a nozzle. Needed is a pump that takes in air and discharges it at higher pressure. That is exactly what the combination of compressor, burner and turbine of a jet engine is: a pump. The mass flow out is almost exactly equal to the mass flow in, but much faster. The exiting flow is also hot by virtue of the heat released in the combustion of a rather small amount of fuel added to the stream. The high temperature of the exhaust means that it has more thermal energy and that helps to create a higher speed jet than it would if the flow was cold.

Explaining how the jet engine works as a pump is not easy without resorting to mathematics and abstract ideas about thermodynamics. It can, however, be described using the perspective of a person who understands the functions of a piston engine like the one that propels an automobile.

The classical piston engine works with a definite mass of air. The motion of the piston within the cylinder causes it to undergo a number of steps that result in the production of useful work, or, if you will, work over time, called power.

The four steps are: 1. intake to fill the cylinder volume, 2. work investment in compressing the air (actually a mixture of air and fuel) and combustion (a rapid deposition of heat into the air) while the cylinder volume is minimum, 3. expansion of the cylinder gas, and 4. exhaust. While these four steps describe a classical four-stroke internal combustion engine, the significant elements of a power budget for the engine involve only compression, combustion, and expansion. Intake and exhaust normally play relatively small roles in an energy accounting.

Compression must be provided either by a starter or as a fraction of the power output by another cylinder via a flywheel. In combustion at minimum volume (initiated by a spark), pressure rises to very high values because the volume during which this very short combustion process (explosion) takes place is, for all practical purposes, fixed. The engineer calls this *constant volume* combustion. The high pressure on the piston exerts a large force, producing a lot of power when the piston is moving to expand the gas volume, some of which is used to compress the air in other cylinders. After expanding to maximum cylinder volume, the hot gas is discharged to the atmosphere (generally at a pressure higher than atmospheric pressure, hence the noise). This batch processing in the piston engine is necessitated by the nature of the combustion of fuel that takes place in a few thousands of a second. Continuous power output from such an engine results from a high rate of such 'batch' events from the many cylinders and the pistons turning at high RPM. These are the steps involved in the internal combustion engine processing 'batches' of air.

The gas turbine engine undergoes the same elementary processes with two significant differences. The first is that the air is handled not in batches, but in a continuous way.

The second attribute of the jet engine that distinguishes it from the piston engine is that combustion takes places in an environment wherein the pressure stays fixed. The net result of combustion is a large increase in *volume*. The engineer calls this *constant pressure* combustion. It is carried out continuously with a steady 'flame'. The large volume flow rate, made larger by virtue of the heat added to the flow, allows the turbine to extract more power from the moving gas than is required by the compressor flow to reach the same pressure. Thus, a net power output is realized.

One way to visualize this is using the understanding of the way in which pistons and cylinders function. The work involved in compression using a piston/cylinder is proportional to the volume swept by the piston, the so-called displacement. One can imagine that heat is added to the compressed gas and its space is kept at the compressed air pressure by a magical piston whose diameter grows as the heat is added. When heat addition is complete, the larger piston can expand and the work obtained from the larger piston is larger (because the displacement is larger) than the work required to compress the air in the compression part of the process. Thus, net work is produced with a constant pressure combustion process. Of course, the gas turbine engine involves neither pistons, nor cylinders, nor magic, but the idea illustrates the point.

After the power needs of the compressor in a jet engine are satisfied, the pressure of the exhaust gas is still high, higher than air pressure at the inlet because the expansion work is larger than the compression work. Thus the three components constitute a pump and therewith the ability to create a thrust-producing jet.

During constant pressure combustion, the volume of heated air is larger when more heat is added and the resultant temperature is higher. This is the key as to why the turbine inlet temperature is so critical to the performance of the gas turbine engine.

Some details

In an aeroplane application, the gas turbine engine has an inlet and a nozzle. These two components operate in opposite ways: the inlet (also called a diffuser) slows the flow while the nozzle accelerates it. There is no involvement of mechanical power or heat to change conditions of the flow through them so that flow along a streamline must obey an energy conservation principle. A form of this idea is termed Bernoulli's principle. The air involved has a certain amount of energy in only two forms, thermal and kinetic energies. Thus, changes in kinetic energy (speed) must be reflected in our normal measure of thermal energy, temperature. In the inlet, the flow slows and the temperature rises. Conversely, in a nozzle flow the gas cools as it proceeds faster out a nozzle to form a jet. In the absence of friction and other thermodynamic not-so-niceties, the temperature changes correlate with pressure changes. In the inlet, the pressure rises and a portion of the compression is done that the compressor will finish. In the nozzle, on the other hand, the pressure falls from the pump output to the lower pressure of the environment.

Digging a little deeper into the mysteries of aeronautics

The generation of lift and thrust by imparting momentum to some of the air in which an aeroplanes flies always involves a boundary between the air involved and the air that is not involved. The two masses of air necessarily move at different speeds. If we take the viewpoint of the boundary, we note that the air environment seems to be rotating. The rotational motion is a vortex.

Vorticity, or vortex motion, is intimately and necessarily associated with flight. The vortex associated with lift comes in the form of a vortex line emanating from the wing tips and stretches rearward. Similarly, trailing vortices issue from the tips of propellers. When a propeller turns sufficiently rapidly relative to the forward motion of the aeroplane and condensation makes the vortices visible, they resemble rings not unlike a series of smoke rings. When the rings are very tightly spaced, they take on the appearance of a sheet. That is how the boundary of jet engine exhaust may be visualized. The energy associated with a difference of velocity on either side of the vortex sheet, *i.e.*, the strength of the vortex, determines how much of the energy produced by the propulsion system is wasted. At a fundamental level, the minimization of that waste is achieved through the use of a turbofan instead of a turbojet for a given amount of thrust. At the heart of the issue is the efficiency of the propulsion system, and that is reflected in the fuel consumption of the jet engine.

An interesting thing about a vortex is that it cannot end in space; it must end on a boundary or form a closed loop. Thus, vortices are always in the form of a ring or other closed form. So where is this loop in the case of a wing? The wing must be part of the picture, and indeed it is. The air motion around the wing behaves as if there is a vortex bound to the wing, meaning flow over the top of the wing is faster than flow along its lower surface. We call this a *bound* vortex and it is necessary for the wing to produce lift. The trailing vortices reach back to the airport where a *starting* vortex is created when the pilot rotates the aeroplane to take off. These elements form a closed loop. Naturally they dissipate, but in theory we have the necessary loop.

There is another dimension of flight that is interesting. One does not normally think of air as being a viscous fluid like oil or honey, but it is, at least that part of the flow immediately rubbing along the surfaces of the wing. The air affected by the friction near the surface is called a *boundary layer*. It can be characterized as a region where friction has locally slowed the air to a lower speed than the air that has not felt the friction from the wall.

If there were such a thing as an inviscid fluid, *i.e.*, one that does not experience the effects of viscosity, then one can readily show (because the equations of motion are very simple) that the flow from the trailing edge of a wing at a modest angle of attack would leave the aerofoil from the upper surface of the aerofoil and fail altogether to form a bound vortex. Thus, air without friction cannot be made to generate lift! It is indeed the inability of the boundary layer air to negotiate the flow around a sharp wing trailing edge that causes the bound vortex to form and thus lift to be realized. Air viscosity is generally not beneficial because it causes an object moving through air to experience drag, but it is also necessary to allow an aerofoil to generate lift. By these means, the trailing edge establishes the flow angle and creates the downward momentum we need to satisfy Newton's laws and obtain lift.

The aviation pioneers recognized this aspect of managing the airflow over their gliders and powered aircraft wings, even though they might not have phrased the explanation in the way stated above. Leading edges of wings or subsonic inlets need to be rounded to allow the air to impact them from various angles. On the other hand, the trailing edges of wings and nozzle exits should be sharp to direct the flow where it is able to generate the momentum in the direction it is needed for lift or thrust. (In supersonic flight, inlets must have sharp lips to slow the air flow down gently through a multitude of weak shock waves.)

Vorticity is the magic of fluid motion and its study makes for challenges. Aviation can't live without it. Even for coffee to be enjoyed with cream or milk, a spoon must provide at least a little vorticity for it to be a café au lait, well mixed!

Bibliography

General references

'Eight Decades of Progress – A Heritage of Aircraft Turbine Technology', GE Aircraft Engines, 1990.

Flight Global Archive, December 1955. www.flightglobal.com/pdfarchive/view/1955/1955%20-%201748.html

Sales brochure Pratt & Whitney JT8D-200, courtesy James Clyne.

Wikipedia and its references for much numerical data about engines, aeroplanes and timeline information.

Archives of the Aircraft Engine Historical Society.

Authored references

Angle, Glenn D., *Aeroplane Engine Encyclopedia*, Otterbein Press Dayton, Ohio, 1921.

Angle, Glenn D. Ed., *Aerosphere 1939 – Including the World's Aircraft Engines with Aircraft Directory*, Aircraft Publications, New York, 1939.

Conner, Margaret, *Hans von Ohain: Elegance in Flight*, American Institute of Aeronautics & Astronautics, Reston, Virginia, 2001

Connors, Jack, et al, New England Air Museum: Digital Recording of a seminar on the History of the Pratt & Whitney J57, ca. 2005.

Connors, Jack, *The Engines of Pratt & Whitney: a Technical History*, American Institute of Aeronautics & Astronautics, Reston, Virginia, 2010.

Conrad, Barnaby, *PanAm – An Aviation Legend*, Council Oak Books, San Francisco, 2013.

Cook, William H., *The Road to the 707*, TYC Publishing, 1991.

Culy, Douglas, 'The Impact of the Engine on the Airframe', Part I, 'The Engines', Torque Meter, *Journal of the Aircraft Engine Historical Society*, Vol. 5, no. 1, Winter 2006.

Flynn, George L., *Escape of the Pacific Clipper*, Brandon Publishing, Boston, 1997, 2016.

Franz, Anselm, *From Jets to Tanks*, AVCO-Lycoming, Stratford Division, 1985.

Galland, Adolf, *The First and the Last*, Bantam War book, 1954, Henry Holt and Company (translated from the German, *Die Ersten und die Letzten*).

von Gersdorff, K. Grasmann, K., *Flugmotoren und Strahltriebwerke*, Bernard & Graefe Verlag, München, 1981, in German.

Golley, John, *Jet – Frank Whittle and the invention of the jet engine*, Datum Publishing, Fulham, UK, 1996.

Heaton, Colin D. and Lewis, Anne-Marie, *The ME 262 Stormbird: From the Pilots Who Flew, Fought, and Survived It,*' Zenith Press, Minneapolis, Minn., 2012.

Heppenheimer, T.A., *Turbulent Skies*, John Wiley and Sons, 1995.

Langston, Lee, S., 'Each Blade a Single Crystal,' *American Scientist*, Vol. 103, No. 1, January 2015.
Ley, Willy, *Engineer's Dream – Projects That Could Come True*, Penguin, 1954.
Makos, Adam, *A Higher Call*, Penguin Group (USA), 2012.
Mellberg, W.F., *Famous Airliners*, Plymouth Press, 1999.
Norris, G. Wagner, M., *Giant Jetliners*, Motorbooks International, Osceola, Wisc., 1997.
Page, Victor, W. *Aviation Engines – Design, Construction, Operation and Repair*, Norman W. Henley Publishing Co, New York, 1919.
Price, Alfred, *The Last Year of the Luftwaffe*, Frontline Books, 2015.
Rummel, R.W., *Howard Hughes and TWA*, Smithsonian Institution Press, 1991.
Schlaifer, R. and Heron, S.D., *Development of Aircraft Engines and Fuels*, Graduate School of Business Administration, Harvard University, Cambridge, Mass., 1950.
Serling, R.J., *Howard Hughes' Airline*, St, Martin's/Marek, 1983.
Smith, George E. and Mindell, David A., 'The Emergence of the Turbofan Engine' in *Atmospheric Flight in the Twentieth Century*, Peter Galison, Alex Rowland, Eds., 2000, Springer Science+Business Media Dordrecht.
Sullivan, Mark P., *Dependable Engines – The story of Pratt & Whitney*, American Institute of Aeronautics and Astronautics, Reston, VA. 2008.
Sutter, Joe, *747: Creating the World's First Jumbo Jet and Other Adventures from Life in Aviation*, Smithsonian Books (Harper-Collins), 2006.
Taylor, C. Fayette, 'Aircraft Propulsion: A review of the Evolution of Aircraft Piston Engines,' *Smithsonian Annals of Flight*, Vol. 1, no. 4, Smithsonian Institution, Washington D.C., 1971.
Tulloch, Judith, *The Aerial Experiment Association – Aviation Pioneers*, Breton Books, Alexander Graham Bell Museum Association, 2014.
Whatley, K.F., 'American Airlines Experience with Turbojet/Turbofan Engines,' ASME Paper 62-GTP-16, March 1962.

Index

The index is given in three parts. The first is an index of the engines made by the various engine manufacturers. These are listed with the engines they produced in an approximate time order. The names of engines and model numbers are used to describe them. When manufacturers produced another's engines under licence, such a relation is noted. Further when the same engine is described by a company name and a military designation, such relation is also noted. Engines that have no manufacturing roots, or a variety of them, are listed in the general index.

The second index relates the manufacturers of aeroplanes to their products.

The third index is the general index where names are either referenced directly or referred to in one or both of the first two indexes.

Engine Index

Allison,
  501, 115
  J33, 103
  J35, 103, 115
  TF41, 157
  V-1710, 73, 143
Armstrong-Siddeley,
  Sapphire, 7, 70
BMW,
  BMW radial, 36–7, 49
  BMW 003, 30, 78, 100
Bristol,
  Hercules, 63, 75
  Jupiter, 30–2
  Olympus, 144
Clerget,
  Rotary, 30
Curtiss,
  OX, 21–2
Curtiss-Wright,
  see Wright
de Havilland,
  Ghost, 125, 127
General Electric,
  CF-6, 86–7, 168–9, 181
  CJ805, 137, 139, 141, 150–2, 156, 165–6, 172
  GE4, 142, 145
  GE90, 170–2, 185
  I-16, 45, 90, 96
  I-40, 45, 95–6
  J35, 95, 103, 109, 119, 124, 141, 143
  J47, 45, 95–7, 103, 119–20, 124, 132, 141, 143–5, 185
  T58, 163
  TF39, 45, 164–6
GE/Allison,
  J33, 95, 102–103, 143
GE/SNECMA,
  CFM56, 169, 181
Hall-Scott,
  Liberty, 23–4, 34, 36, 73
IAE,
  V-2500, 169, 195
Junkers,
  Jumo 004, 70, 78–81, 86, 143–5
Lycoming,
  ALF502, 173–4
  T53, 45, 133, 163–4, 174
  T55, 76, 163–4, 173–4
Metropolitan-Vickers, Metrovick F.2, 70
Oberursel, Rotary, 26–9
Packard,
  Liberty, 24, 32
  V-1650, 47
Pratt&Whitney,
  Hornet, 36, 38, 42, 49, 59
  J42, 103, 134, 141–2
  J48, 103, 134, 141–2
  J52, 142, 157
  J57, 131–2, 134, 137, 141–4, 185
  JT3, 132, 134, 137–8, 142, 144
  JT4, 137–8, 142
  JT8D, 114, 152, 157–60, 179, 181, 185, 192–5
  JT9D, 156, 167–9, 180, 185
  T34, 108
  Wasp, 34–6, 48–50, 59, 63, 72–3, 106, 111–12
Rolls-Royce,
  Avon, 70, 128–9, 141, 152
  Conway, 147–8, 150, 154–7, 160
  Dart, 114
  Derwent, 90, 129, 141, 185
  Merlin, 47
  Nene, 90–2, 103, 114, 141
  RB.203, 170
  RB.211, 166, 168–9, 181
  Spey, 157, 168
  Tay, 141
  Welland, 90, 141
Rolls-Royce/SNECMA,
  Olympus, 142, 144
SNECMA,
  see Rolls-Royce or GE
Westinghouse,
  J30, 97–8, 102–103, 134, 142
Wright,
  Cyclone, 36, 46, 48–51, 61–2, 72–3, 109, 111, 113
  J65, 7, 70, 103, 133
  Simoon, 33–4

Aircraft Index

Airbus,
  A300, 166, 181–2
  A320, 195
AVRO,
  Lancaster, 47
Avro of Canada,
  C-102, 129
Bell Aircraft,
  P-39, 44, 47
  P-59, 90, 93
Boeing,
  707, 101, 110, 113, 134–9, 142, 147–8, 151–2, 154–6, 167, 177, 181, 189–90
  720, 113, 139, 151, 179–81, 184
  727, 152, 157–8, 160, 176, 181
  737, 123, 160, 169, 179–81
  737 MAX, 195–7
  747, 156, 161, 165–9, 177, 179–82
  787, 30, 177, 182, 190
  B-17, 40–1, 46, 49, 56, 72–3, 75, 88
  B-29, 49, 72, 77, 106, 177
  B-47, 95, 101–102, 107, 119–25, 129–31, 137, 141, 145, 181, 189
  B-50, 72–3
  B-52, 123, 129, 131, 134, 142
  Dash 80, 110, 129–30, 134–5
  KC-97, 129–30, 134
  KC-135, 110
  Model 40, 24, 34, 36, 42
  Model 80, 38, 42
  Model 247, 43, 49–51, 53–4, 56, 73
  Model 307, 55–6, 61, 71, 73, 107, 126
  Model 377, 57, 72–3, 77, 106–107, 109, 113, 129–30, 167, 177
British Aerospace,
  111, 157
  Concorde, 30, 61, 66, 122, 129, 142, 152
Consolidated,
  B-24, 49, 73, 75, 77
  XB-46, 94, 119
Convair,
  880, 137–9, 150, 152
  990, 137, 139, 148–52, 156
  B-36, 72–3, 106, 110, 131, 181
  B-58, 137, 178–9
  XC-99, 106
Curtiss,
  JN-4 (Jenny), 17, 22

de Havilland,
  Comet (DH-106), 125–9
  DH-95, 71
  Dragon, 42
Douglas,
  B-18, 56
  DC-2, 43, 50–1, 53–4, 56, 73
  DC-3, 46, 48, 51, 54, 56–7, 73, 111–12
  DC-4, 48, 77, 104, 107
  DC-5, 104
  DC-6, 72, 104, 106–107
  DC-7, 72, 106–109
  DC-8, 135, 137, 142, 148, 151–4, 156, 177
  DC-9, 153–4, 160, 169
  DC-10, 166, 168–9, 181–3
Fokker,
  F.VII, 21, 42
  F-27, 114
  VFW-Fokker,
    614, 173
Ford,
  4-AT Trimotor, 42–3, 47, 51
Gloster,
  E.28/39, 68
  F.1 Meteor, 70, 90, 92
Heinkel,
  He-178, 68
Hughes,
  H-4 (Spruce Goose), 72, 106, 140, 181
Junkers,
  Ju52, 20, 36, 49–50, 56
Lockheed,
  C-5, 164–5
  Constellation, 72, 104–105, 108–109, 122, 135
  Electra, 51, 54, 56, 115, 136–7
  L-1011, 166, 168–9, 181–3
  P-38, 44, 47
  P-80, 92
Martin,
  2-0-2, 111
  4-0-4, 72, 111
  M-130, 59–60
  XB-48, 119
Messerschmitt,
  Me-262, 64, 81, 88–9
Mikoyan-Gurevich,
  MiG-15, 91, 103

MiG-17, 103
North American,
  B-45, 94, 119
  F-100, 134, 178–9
  P-51, 44, 47
Sikorsky,
  S-21, 18
  S-42, 59–60
  S-55, 117–18, 162
  S-58, 117–18, 162
  S-61, 162–3

Sud,
  Caravelle, 152
Sud/Aerospatiale,
  Concorde, 30, 61, 66, 122, 129, 142, 152
Tupolev,
  Tu-95, 100–101, 115, 134
  Tu-104, 125, 129
  Tu-114, 100, 110, 116
  Tu-144, 129
Vickers,
  VC-10, 154
  Viscount, 114

General Index

Aerial Experiment Association (AEA), 15–17, 21
Aeroflot, 116, 125, 153
Aerospatiale, 152
afterburning, 80, 95, 122, 142
aileron, 15–16
air force (various), 19, 31, 88, 90, 95, 101–103, 110, 120, 129–31
air mail, 34–5, 50, 54
Allen, B., 13
Allison, 47, 95, 143
aluminum, 13–14, 20–1, 30, 38, 52, 140, 174
American Airlines, 51, 107, 113, 137, 148, 150, 152, 156, 190
Armstrong-Siddeley, 7, 31, 70
aspect ratio, 13, 95, 175, 177
atmosphere, 3–4, 28, 41
avgas, 57
axial flow compressor, 39, 66–7, 69–70, 92, 95, 119, 123, 128, 133, 143, 163, 173

Bell, A.G., 14–16
biplane, 15, 37–8
bleed air, 77, 178
blow-in doors, 148, 179–80
BOAC, 125
Boeing Air Transport, 34–5, 42, 53
Boeing, W., 18, 34, 41, 54, 104
Brown-Boveri, 39
Büschi, A., 40
Busemann, A., 64, 101
bypass, 78, 83, 147, 149–50, 153, 155–72, 175, 179, 181, 184–5

CAB, 135
cabin pressure, 56, 126

castor oil, 28
Clipper, 59–62, 135, 167
Collier Trophy, 52, 134
combustor, 82–4, 91, 97, 132, 158–9, 174
compressor stall, 70, 79, 81, 126, 134, 167
compressor, radial flow, 40, 69–70, 91, 96
Consolidated, 8, 49, 58, 77
Convair, 135–9, 148–52, 156
Cook, W.H., 101, 119, 189
cowl, 28–9, 50–1, 156, 179
Curtiss-Wright, 18, 32, 41, 112
Curtiss, G.H., 15–18, 21, 32

Daimler, xvii, 24, 65, 78, 82
Decher, S.H., 80, 164
de Havilland, 42, 71, 125–8
Delta Airlines, 137–8, 152, 182
diffuser, 66, 79, 83, 132
directional solidification, 85
dirigible, 10–11, 65
Doolittle, J.H., 57–8
Douglas, D.W., 18, 153
drag, xv, 4–5, 37–9, 64–5, 150, 175, 177, 187–9
Dumont, A.S., 11

Earhart, E., 19, 21, 51
engine-airframe integration, 178, 191–2, 195

fan blade, 150, 165, 167, 170–1
flame holder, 83
flaps, 59, 122, 135, 176, 180
flying boat, 17, 58–61, 71, 117, 140, 162
Franz, A., 78, 102
Frye, J., 50

gas generator, 76, 166, 174
gearing, 46, 56, 111, 172–4
General Dynamics, 137, 152
Giffard, J., 10
Griffith, A.A., 39, 67, 70
Grumman, 92, 103, 141
Guggenheim, 21, 64

Heinkel, 68, 78
helicopter, 76, 102, 117–18, 162–3, 173
Hispano Suiza, 21, 23–4
Hornet, 36, 38, 42, 49, 59
Hughes, H.R., 71–2, 108–10, 135–40, 150, 152

induced drag, 38, 175, 177
inlet, 170, 178–80, 185, 195, 208, 210
inlet guide vane, 92, 144–6, 165, 170
intercooling, 41

Johnson, C., 104
Junkers, 42–3, 53, 64, 71, 78, 99, 102, 163
Junkers, H., 19–20, 43

KLM, 19

labels (engine names), 45, 162
Langley, S.P., 11, 13, 15
Lawrance, C.L., 30–1, 33
Ley, W., 104
Lindbergh, C., 19, 21, 31, 56, 59
Lufthansa, 20, 49–50
Lycoming, 31, 76, 78, 102, 163, 173–4

Mach number, 116, 137, 142, 150, 174, 179, 188–9
Mach, E., 5
Martin, 16, 18, 21, 23, 58–60, 72, 111, 119, 122
Martin, G.L., 8, 16, 18, 58, 111, 119
Maxim, H., 11
Mead, G., 23, 33
mixer, 154, 159, 190
monoplane, 20, 38, 42, 59
Moss, S.A., 40

NACA, 40, 51, 64, 101
NASA, 64, 101
Navy (U.S.), 30, 33, 62, 97–8, 102–103, 108, 115, 141–2, 162–3

Neumann, G., 146, 165–6
Newton, I., 188
noise, 57, 116, 122, 136, 147–8, 154–5, 158–60, 166, 169, 178–9, 189–90
Northrop, J.K., 19, 53–4
nozzle, 69, 79–81, 83–7, 136, 147–8, 154–5, 178, 189, 206, 208

octane, 57–8, 63, 106
Oestrich, H., 30, 78
ovalization, 167

Pan American Airways, 6, 56, 58–9, 61–2, 104, 135, 167
Paperclip (Operation), 100
parasitic drag, 38
Piasecki, F., 117
pitch, 52–3, 59, 122, 160, 173, 189
Porter, R., 10
Pratt & Whitney, 18, 32
pressure, 3–5, 79–84, 126, 132–4, 143–4, 169–70, 188, 201, 206
pressure ratio, 66–7, 81, 95, 132, 143, 158–60, 169
propellers, 52
propfan, 174

radial engine, 30
range, 199–201
Rateau, A.C.F., 40
Rentschler, F., 23, 32–3, 41, 102
reverser, 122, 154, 160, 162, 165
RLM, 78
rocket, 120–2, 125–6
rotary engine, 13, 17, 23–30, 32, 37, 73, 117
Rummel, R., 108, 135–6, 138–40

Safran (see also SNECMA), 7, 78, 169
Schairer, G., 101, 119
Seguin brothers, 26
Selfridge, T., 15–16
shock wave, 5, 66, 149–50, 160, 188–9
Sikorsky, 17, 41, 54, 58–61, 117–18, 162–3
slats, 176
SNECMA, 7, 30, 78, 141–2, 169, 173
spoilers, 123
spool, 80, 132, 137, 144, 146, 185
stall, 52, 67, 70, 78–9, 81, 88, 122, 126, 134, 146, 167, 179, 189
stator, 66–7, 78–9, 144–6, 165, 170, 173–4

Stout, W., 43
supercharge, 8, 39, 41, 46, 62–3, 74, 95, 111
supersonic, 66, 122, 129, 137, 142, 145, 152, 170, 178, 188
Sutter, J., 167–8

Taylor, C., 12
Taylor, C.F., 31, 111
TCA, 129
thrust reverser, 122, 154, 160, 165
time between overhaul, 81, 112–14
tip speed, 66, 116, 172
Toolco, 136–9
Transcontinental and Western Airline, 35, 50, 53, 71, 104
trijets, 182
Trimotor, 20, 42–3, 47, 51
Trippe, J.T., 59, 138, 167
turbine, 39–40, 66–9, 84, 87, 144
turbocharger, 8, 40–1, 84, 95
turbojet, 67, 78, 125, 131–5, 141–4, 147–8
turboprop, 76, 100, 114–16
turboshaft, 76, 102, 111, 162–3, 173–4
TWA (Transworld Airlines), xvi, 104, 106, 108–10, 135–40, 150, 152

UATC, 41, 54
United Airlines, 35, 51–4, 152
USAAC, 54, 56–7

Vickers, 47, 70, 110, 112, 114, 125, 154
von Karman, T., 70, 101
von Ohain, H.P., 67–8, 78, 82, 102
vortex, 38, 83, 123, 177, 189–90
vortex generator, 123, 189–90

water injection, 72, 75, 120, 122, 139, 145
weights, 73, 184–5, 198–200
Whittle, F., 67–70, 81–2, 90, 147
wind tunnel, 12, 64, 101, 119, 191
wing sweep, 89, 101, 103
winglet, 123, 177
wings, 12, 37–9, 64–5, 101, 119, 122–3, 175, 177
Wright brothers, 11–18

Zeppelin, 65